An Exploratorium Book

TRACES OF TIME

THE BEAUTY OF CHANGE IN NATURE

by Pat Murphy and Paul Doherty

Photographs by William Neill

Foreword by Diane Ackerman

CHRONICLE BOOKS

SAN FRANCISCO

Library of Congress Cataloging-in-Publication Data:

Murphy, Pat.

Traces of Time: The beauty of change in nature / by Pat Murphy and Paul Doherty; photographs by William Neill; foreword by Diane Ackerman.

p. cm.

Includes bibliographical references and index.

ISBN 0-8118-2857-3 (pb)

1. Geology—History 2. Evolution (Biology) 3. Science—Philosophy. I. Doherty, Paul. II. Neill, William. III. Title.

QE11. M87 2000

508—dc21   99-087677

Front cover photos (top, left to right): Sunflowers and Phototropism; Foxtail Grasses; Half Dome and Tenaya Canyon, Yosemite National Park, California; Obsidian, Mono Lake, California; (bottom) Moving Rocks, Mojave Desert, California.
Back cover photo: Mount Conness and Ellery Lake, Inyo National Forest, California.

Designed and typeset by Open, New York

Printed in Hong Kong

Distributed in Canada by Raincoast Books

9050 Shaughnessy Street

Vancouver, British Columbia V6P 6E5

10 9 8 7 6 5 4 3 2 1

Chronicle Books, LLC

85 Second Street

San Francisco, California 94105

www.chroniclebooks.com

Fossil Dragonfly Larva, Dating from 20 Million Years Ago
(page 2)

For anyone
who has ever considered
a grain of sand
and imagined the boulder
where it began.
—Pat Murphy and Paul Doherty

For my wife Sadhna,
for showing me the meaning of love,
and my daughter Tara,
for all the magic and wonder
she brings to my life.
—William Neill

CONTENTS

FOREWORD by Diane Ackerman

Time is no more invisible than an elk drinking at the river—it sends ripples across the water, carves footprints in the mud, leaves rings at the heart of a tree. Nature offers sundials and calendars at every turn.

As I write this, it is late summer in North America. Bicycling today through the rolling hills of upstate New York, I noticed Queen Anne's lace dotting the roadsides along with blue chicory, spiky brown teasels, and purple loosestrife. The wild sweetpeas that used to tangle with any available tree, bush, or lawn ornament have withered. So, too, have svelte pink dame's rocket and common daylilies. Every day for weeks I seemed to disturb one female ring-necked killdeer, a stilty-legged low-nesting bird notorious for distracting attention from its brood by dragging a wing as it scurries away. Yesterday and today, no killdeer. Its chicks have flown, as they do late in the season.

Clouds of fireflies were June's festival of lights. Now the moonless nights are dark and fiercely raucous. Cicadas sound like a thousand snakes rattling in unison. Sometimes a raccoon will butcher a baby rabbit, whose high-pitched screams are spine-wrenching to hear, even though I know raccoons have babies to feed, too. Two generations of house wrens have laboriously nested in bird boxes hanging from a sycamore, fed their ever-peeping chicks, and flown. The fifty-year-old magnolia outside my bay window has formed thick ropy buds that next spring will open into brandy snifter–shaped flowers.

In early summer, the brown corduroy of plowed fields smelled like freshly turned moss, manure, worms, and cinnamon. Risen corn now means the season is waning. Somehow the cornstalks always surprise me at summer's end. But I do love corn's golden tassels catching the breeze like wayward forelocks. All the cornstalks seem to be pointing up, and remind me of Leonardo da Vinci's portrait of John the Baptist with one hand raised, finger pointing toward heaven. By August the fields will become armies of withered scarecrows, and I'll fret as I bicycle past them, fleet over the earth but disturbed by the fleeing season.

I also respond to the more ethereal traces of time, such as sunlight prowling around the house and climbing in the west-facing bay window at around 3:00 P.M. Best to let the filmy, semi-permeable blind down then, lest our local star bleach the cushions. Driving through mountain passes, one can see solid time laid out in strata, geologic eras piled like so many oriental rugs.

Nature's calendar depends on locale, of course. A crust of ice on a lake in the East signals winter, but not so in the Klondike. In the eastern United States, migrating birds draw check marks across the sky in autumn while squirrels bury nuts

Cracked Mud on River Rocks, Green River, Utah
(opposite)

9

and grow fluffy bellies to prepare for winter. But in California monarch butterflies will be invading eucalyptus groves.

One misty cold morning recently, I strolled along the gorge trail to Taughannock Falls. The waters were high and violent, and when I crossed a bridge to the lookout point near the bottom of the towering falls, the air grew thick with spray and aerosols of mud and silt. Above, time revealed itself in ribbons of rock. The falls spilled over a cauldron of limestone, the hard core of eons. But where I stood the canyon was mainly shale, a soft graphitelike rock that is constantly crumbling, powdering, breaking down. Two contrasting types of rock frame this gorge. When I stand at the base of them, I can see clearly, as one rarely can in life, a basic principle of time in nature—what resists and what falls away. The shale begins with water and ends with water in a cycle of construction and destruction. The woods, too, are constantly sloughing. I can see it in the peeling bark, the decaying logs, the crumbling stone. In that setting, I recalled a friend who had recently died, an ebullient young woman whose foot cancer ultimately killed her. To the end she stayed upbeat, positive, determined to live. She became engaged and married, resisting the idea of her death while her body fell away.

Time both races and pools in one's memory. I wonder: do the same neurons convey both illusions? When I was in junior high school in Allentown, Pennsylvania, every classroom had a sign beneath a large wall clock that read: "Time passes, will you?" We students understood that the question was meant to motivate, despite its dubi-ous grammar. Most of us would pass junior high school, but what we didn't know then—when we were wont to stash a dead groundhog in a locker before the weekend, or insert Popsicle sticks between the strings of the music teacher's piano, or brew 100-proof alcohol in the chemistry lab—was that time would pass faster than we wished, faster than our brains could reckon, as fast as fatal rocketry. Some students would die in the Vietnam War, and others fall victim to a mysterious virus or bacterium. But at the time all we thought about was growing up, and not gradually, but in a single breakthrough of height, breasts, and vocal cords. We longed for the fine lines of worldliness visible in the close-ups of glamorous movie stars. We traced our stages of growth with sweet sixteen parties, bar mitzvahs, proms, and other rites of passage. That still happens, of course, as kids move from a jump rope to a rope of pearls.

Meanwhile, time leaves a trail of fossils in limestone. Meteor showers guide us through the seasons, as does the waltz of the constellations. Each month a curvaceous moon fattens and skinnies. Lengthening shadows alert us to day's end as surely as sundials. If we're lucky, a long-haired comet may appear for the first time in 4,000 years. When we peer at exquisite nature photographs, time stops for precious moments, allowing us to step between tock and tick, while a camera gasps.

Burnt Trees and Shadows on Snow, Blacktail Plateau, Yellowstone National Park, Wyoming (opposite)

10

INTRODUCTION by Pat Murphy

The simple act of seeing is not as simple as it seems.

I used to think I saw the world as it was. After working at San Francisco's Exploratorium more than seventeen years, I have realized that isn't exactly true. While working with optical illusions, while writing about science and nature, while working with people who were constantly opening my eyes to phenomena that I had previously overlooked, I realized something significant. Most of the time, I wasn't really seeing the world as it was. Instead, I was seeing what I had learned to see. When I learned something new, I saw a different world.

I'll give you an example. Many years ago, I spent a summer at an archaeological dig just north of Arizona's Meteor Crater. On my first walk with one of the archaeologists, I glanced at the ground and saw only dirt. The archaeologist glanced down at the exact same spot and immediately picked up a bit of broken pottery and a piece of stone that had been worked by the Anasazi people who had lived in this place long ago. I was amazed—how could she spot these things in the jumble of dirt and rocks?

After a month at the dig (a month spent grubbing in the dirt and unearthing ancient artifacts), I could stroll down a trail and spot the bits of pottery and worked stone. Having learned what these things looked like, I could see them, even though they had been invisible to me before.

This shift in perception happens on both the small scale and the large scale. On that same dig, I went out on survey, leaving the main archaeological site and searching the surrounding hills for outlying settlements. Again, I had to learn what to look for. I learned to spot stones that had been laid in a straight line, an unnatural arrangement indicating where a wall had once stood. I learned to look for certain non-native plants, which revealed that a patch of earth had once been a garden. The signs were obvious once they were pointed out— but invisible until then.

I went on that archaeological dig just before I started working at the Exploratorium, and the museum reinforced the lesson I had learned on the dig. San Francisco's Exploratorium is a museum of

12

science, art, and human perception. The museum is filled with exhibits that allow people to explore and experiment. Visitors can study the patterns made by vibrating pins, play with a captive tornado, experiment with optical illusions, and learn the limitations of their own visual perception.

Some people think of the Exploratorium as a science museum, since many of our exhibits explain how the world works. Others think of the Exploratorium as an art museum, since artists in our artist-in-residence program have produced some of the museum's most spectacular exhibits. Still others think of it as a children's museum, since children have a great time exploring and experimenting here.

I usually think of the Exploratorium as a museum dedicated to helping people notice things that they usually overlook. The signs on our exhibits tell visitors about things "to do and notice," and noticing is a big part of what goes on here. The Exploratorium is a museum of discovery—not just other people's discoveries, but your own personal discoveries. The Exploratorium is a place designed to shift your perception of the world.

The perceptual shift that led to this book began with a piece of petrified wood. I was leading a tour of the Southwest designed to open the participants' eyes to the natural world. As so often happens in the course of teaching, I was the one who learned.

The tour began in Utah's Park City, a ski-resort town that didn't have much to offer in the way of nature walks. So I began on the first morning by sitting with my group and passing around a piece of petrified wood, purchased in an Arizona

Growth Rings of a Petrified Log

Petrified wood preserves a record of growing conditions millions of years ago. Growth rings, like the ones visible in this fossil log, are most clearly defined in trees that grow in temperate climates, away from the equator. If a tree lacks growth rings, chances are it grew in the tropics, where seasonal variations are minimal.

In Europe, fossil woods dating from the Permian Period (280 to 225 million years ago) generally lack growth rings. Back then, according to the theory of plate tectonics, Europe was in the tropics.

rock shop. "Look at this carefully," I told them. "What do you see?"

There is no right answer to that question. The people on the tour had a variety of answers. The answer that I gave, the answer that was true for me, can be summed up in a single word: time. For me, that piece of petrified wood was a message from the past.

That chunk of petrified wood—a tree that was turned to stone—has a story to tell. The story began some 225 million years ago, when Arizona's Petrified Forest National Park and the surrounding land was a floodplain. Tree-covered islands rose above the marshy ground. Then the blast of a volcanic eruption shook the earth, leveled the trees, and covered the fallen debris with layers of volcanic ash. Rains came and flooded the area, dissolving the volcanic ash and the minerals it contained.

One of those minerals was silica, or quartz, the most abundant mineral on the Earth's surface. Over the passing years, that dissolved silica crystallized in the cells of the fallen trees, replacing the organic matter bit by bit. Iron, manganese, and other minerals colored the silica red and black and white and yellow.

Many many years later, a rock hound found that piece of petrified wood and sold it to a rock shop. The rock shop owner sliced across what was once the tree's trunk and polished the cut surface smooth.

In this piece of petrified wood, you can still see the original structure of the tree, the growth rings that were laid down when the tree was alive. The width of each growth ring reveals how much

**River Rapids,
Short Exposure Time**
(left)

**River Rapids,
Long Exposure Time**
(right)

Your perception of time affects how you see the world. The photograph on the left was taken with an exposure of one-thirtieth of a second, a little bit faster than human eyes process light. At this speed, some of the water droplets are frozen in place. The photograph on the right was taken with an exposure of half a second, much slower than human eyes process light. In this view, you see the path of each water drop as a strand of white. If you were to look at these rapids, you would see something in between these two views.

the tree grew in a particular year. If a year was good—with lots of water and good growing conditions—the tree grew more and the ring is wider. These growth rings reveal what the weather was like when that tree was growing, hundreds of millions of years ago.

That petrified wood started me thinking about time and how its passage is recorded by the natural world. On that same trip to the Southwest, I traveled to Utah's Goblin Valley, drove winding dirt roads through Capital Reef National Park, hiked in a slot canyon in Grand Staircase–Escalante National Monument, camped in Monument Valley, and visited the rim of the Grand Canyon. Along the way, I became increasingly aware of time and how it leaves traces of its passage in the landscape.

I gazed at the multicolored walls of the Grand Canyon, considering layer upon layer of sedimentary rock. The layers of stone chart the history of this place—showing when the sea filled the basin and when it left, when volcanoes spewed ash and lava across the landscape, when the land was pushed up by movements of the Earth's crust, when it was shaken by earthquakes. Layers of geological history dating back two billion years are there for people to read. The history is preserved in the stone.

If something is permanent and unchanging, people say it's as solid as a rock. What a strange saying that is. The rocks of the Grand Canyon are solid—but they are neither permanent nor unchanging. The very presence of the canyon testifies to that. To open up this dramatic vista, the waters of the Colorado River have washed away layer upon layer of rock. It is the absence

of rock where rock once was that makes this view so amazing.

But it doesn't take a visit to the Grand Canyon to see that rocks change. Study the rocks on an ocean beach and you'll find jagged ones and smooth ones shaped by the waves and ones that are halfway there. You may see grains of sand—tiny fragments of rock—and chunks of sandstone that are made of many sand grains stuck together. The sand is on its way to becoming sandstone and the sandstone is wearing down to make sand again.

Rocks are always changing—but their change takes place on a different time scale than we humans are used to observing. We humans have a limited view of time. We note the passage of days, of months, of years. A century is more than a lifetime; by human reckoning, that's a long time. When we measure by human standards, a thousand years is a very long time.

Measured by other standards, a year is inconsequential; a decade is fleeting; a century barely registers. In writing this book, I became increasingly aware of changes that occur too slowly for humans to observe. My research opened my eyes to marks left by events that happened a million years ago and made me aware of the slow changes going on around me every day.

To write this book, I teamed up with one of the Exploratorium's physicists, Paul Doherty. Paul is a rock climber with a passion for geology, a spirit of adventure, and a deep understanding of the natural world. When Paul and I began talking about the book's organization, we decided that each chapter should deal with a different time scale. Each chapter tackles a time frame 100 times that of the preceding chapter. We begin with years, then move to hundreds of years, tens of thousands of years (100 x 100), and millions of years (100 x 100 x 100).

With this book, we wanted to show you some of the ways that time leaves its marks on the world; we wanted to help you see the traces that time left behind. To that end, we brought in William Neill, a noted nature photographer whose work we admire. Looking through Bill's extensive archives of beautiful nature photographs, we found many images that revealed the passage of time.

As Paul and I worked on this book, we would look at the photographs and tell each other stories. My background is in biology and my stories tended to be short ones—stories about years and centuries. I pointed out the sharp crook in the trunk of a lodgepole pine, the legacy of a winter of heavy snowfall when the tree was a sapling. I considered a photograph of Yosemite's Lembert Dome, and talked about the creeping advance of green plants over the smooth granite, the slow change of biological succession.

Paul, with his background in geology, focused on longer stories—ones that took place over thousands or millions of years. He looked at that same photograph of Lembert Dome and saw the footprints of a glacier. The ice had melted 10,000 years ago, but its passage was preserved in the shape of the rock. He talked about young mountains—and he meant that they were just a few million years old.

"What do you see?" Paul and I asked each other as we looked at photographs. "You say that you know a glacier was here 10,000 years ago, but what do you see that tells you that?"

In captions and text, we've written down what we see in these pictures. We've told the stories about what we see and the stories about what others have seen.

In the end, this is a book of stories. Scientists, like all human beings, are storytellers. The stories scientists tell are called theories, but that's just a name. Scientists observe the world and try to make sense of it. Like people everywhere, scientists tell stories that explain the world as they see it.

We haven't told all the stories that can be seen in these pictures, and these pictures do not include all the places in the world where you can see traces that show the passage of time. This is just the beginning. We want to open your eyes to the traces of time—and send you out to make your own discoveries. Look in the world around you and discover other places where time has left its mark. Happy hunting!

On the table of contents page, you may have noticed four patterns of vertical lines. In this book, these patterns represent the different periods of time by using exponentials—in this case the powers of ten—to determine the spacing between the lines. For the relatively short span of year-by-year events we have represented a ten-year span ($10 = 10^1$), and have separated the lines for these events by one unit of space ($1/8"$). For lines representing hundreds of years ($100 = 10^2$) we have separated the marks by two spaces; for tens of thousands of years ($10,000 = 10^4$), four spaces; and for millions of years ($1,000,000 = 10^6$), six spaces. If we had used a linear approach to represent the spacing, and the lines for years were $1/150^{th}$ of an inch apart, the lines for millions of years would have to be 1,500 inches—or more than 180 feet—apart.

16

Mount Conness and Ellery Lake, Inyo National Forest, California

In a single scene, you can see traces of events that occurred in the last year, the last century, the last 10,000 years, and the last million years.

The light-colored granite of Mount Conness, the peak just to the right of the center of the photograph, formed when a pool of molten magma cooled underground millions of years ago. This rock was pushed to the Earth's surface by the great movements of the drifting plates that make up the Earth's crust, then was exposed by erosion.

The granite was sculpted during the last ice age, when much of North America was covered by glaciers. Study Mount Conness and you'll see a narrow ridge, extending toward the camera. On either side of the ridge is a bowl, or cirque, carved by a glacier that melted at the end of the last ice age, about 10,000 years ago.

Consider the trees that cling to the rocky flanks of the nearby hills. Ten thousand years ago, glaciers scraped this land down to bare rock. Over the passing centuries, wind and weather have eroded this rock, creating pockets of soil where plants have taken hold. The trees that you see are part of an ongoing process in which one plant community is replaced by another, a transformation that occurs in a time span of hundreds of years.

Finally, take a look at the ice on Ellery Lake. The gaps in the snowy surface indicate that the ice is breaking up, a sign of spring, a seasonal change that happens each year.

YEAR BY YEAR

The natural world has stories to tell. Becoming aware of nature's stories is a little like learning to read a foreign script. People who devote their lives to studying nature learn to notice punctuation marks that nature inserts in her writing. They learn to interpret the traces left behind by natural processes. They learn to pay attention to indicators that others overlook.

Reading the past begins with reading the present. Back in 1785, James Hutton, a Scottish naturalist who came to be known as the father of modern geology, wrote "the present is the key to the past."

By noticing and understanding how the natural world changes from day to day, from month to month, and from year to year, you can come to read the messages from hundreds of years ago, thousands of years ago, even millions of years ago. To read these messages from nature, you must learn to see changes that others overlook, learn to notice details that others ignore.

Next time you are hiking, look for places where trees have fallen across the trail. Trail crews saw through these trees to open the path. When Paul Doherty, one of the authors of this book, is skiing through the woods, he uses these cleanly sawn ends to find his way, knowing that the trail runs between them.

But these sawn trunks reveal another path, a path through time. In a temperate climate, each concentric ring in the cross section of a fallen tree marks the passage of a year. During the growing season, the cambium layer just under the tree's bark produces new cells. Early in the season, these cells are large, with thin walls. Late in the season, the new cells are smaller, with thicker walls. The growth rings that you see mark the difference between early growth and late growth. By counting the rings, you can determine how old a tree was when it fell.

That's a story you've probably heard before, one that many of us learned in elementary school. But that's just the beginning of the story told by the rings of a tree. Tree rings reveal much more than a tree's age. The rings of the tree stump pictured on pages 18–19 tell of easy living. These rings are fairly evenly spaced, an indication that the tree grew in an area where the soil was good and water was abundant.

Trees growing under harsher conditions, where water is lacking in some years and abundant in others, have rings of varying width. A wide growth ring indicates a year with good growing conditions; a narrow ring indicates the opposite. The rings of the fir tree shown on page 25 reveal that this tree grew where conditions changed from year to year.

Tree rings preserve the record of many events.

20

Growth Rings of a Tree Cut Down in 1990
(previous page)

**Horsetail Falls at Sunset,
Yosemite National Park,
California** (above)

This dramatic photograph of Horsetail Falls was taken in late February, the only time of year when a photograph like this is possible. For just a few weeks in the year, light from the setting sun shines through a gap in the surrounding cliffs to illuminate the torrent of falling water. During the rest of the year, the falls are shaded from sunset light.

**Boulders at Sunset,
4:31 P.M.** (top)
5:01 P.M. (middle)
5:37 P.M. (bottom)

Observing shadows can make you more aware of the sun's daily movements. Notice the shadows of the boulders in these photographs taken near sunset at intervals of about half an hour. In just half an hour, the shadows change dramatically.

Next time you're outside in the middle of a sunny day, watch the shadows around you. Over the course of an hour, notice which way the shadows move. If you live in the northern hemisphere, the shadows will move in the direction that we call clockwise. In the southern hemisphere, the shadows travel the other way.

Before the invention of mechanical clocks, sundials tracked the hours. The inventors of mechanical clocks lived in the northern hemisphere. When they set the direction of the clock's hands, they mimicked the movement of the shadow on the sundial. If the inventors had lived in Australia or South America, clockwise might be the other way around.

Fire sweeps through an area, scarring a tree and leaving its mark in the tree's rings. A late-spring frost freezes water in the tree's growing cells, causing them to burst and leaving a whitish ring in the wood. A porcupine gnaws through a tree's bark and sapwood; an invasion of insects attacks a grove of trees; a heavy snowfall bends a tree beneath it; a volcanic eruption sends dust high into the atmosphere, reducing the sunlight and making summer weather cooler than usual. All these events leave traces in tree rings, traces that a trained observer can read.

Dendrochronologists are scientists who study tree rings and use that information to determine dates and past conditions. The word comes from the Greek words related to trees (*dendrites*) and time (*chronos*).

The science of dendrochronology got its start in 1901 when astronomer Andrew Ellicott Douglass began searching for a natural record that might reveal ancient cycles in sunspot activity. He began investigating tree rings, reasoning that variations in sunspot activity would affect climate and that these climatic changes would be recorded in the tree rings. At lumberyards in Flagstaff, Arizona, he measured the widths of tree rings and established the close association between ring width and rainfall.

In core samples taken from living trees, Douglass looked for identifiable patterns of wide and narrow rings and noted when each ring formed. He then examined the ring patterns in dead standing trees, searching for sections that overlapped those of the living trees. Since he knew each ring represented a year, he could work

Changes in the Sun's Position Over a Year
(opposite)

As the Earth orbits the sun, the time and position of the sunrise and sunset change. The length of the day and the angle of the sun at any time of day vary with the season over the course of the year.

This remarkable photograph, taken in the late 1970s, documents this annual cycle. Astrophotographer Dennis di Cicco fixed a camera in position outside his house in Massachusetts. Over a year, he exposed the film once a week at the same time of day. The resulting photograph shows the analemma, the shape traced by the sun's movement over the course of a year. The sun is highest in the sky in the summer and lowest in the winter. Dennis di Cicco/ *Sky & Telescope* magazine

Sunflowers and Phototropism
(above)

Green plants move over the course of the day to maximize the sunlight that shines on their leaves, a tendency known as phototropism. The sunflowers in Pat Murphy's garden follow the sun each day, turning to face it in the morning and moving to follow it until the house blocks the light.

backward from the overlapping sections, assigning a year to each ring. To extend his time line back even farther, Douglass examined wood from beams found in the ruins of prehistoric cliff and mesa dwellings in the area.

By 1928, Douglass had developed two separate continuous patterns of tree rings or chronologies, each covering more than 500 years. One chronology started with living trees and worked backward. In this chronology, each ring could be assigned to a specific year. The second chronology was based on prehistoric beams. The rings could not be assigned to actual dates, and Douglass had not found any overlaps between the chronologies. Archaeologists estimated that the gap was at least 500 years.

In 1929, a single sample from ruins at Showlow, Arizona, allowed Douglass to bridge the gap between his chronologies. The gap turned out to be less than 100 years. Using this chronology, he could establish precisely when a particular tree was felled. Analyzing the timbers used in constructing Pueblo Bonito, a prehistoric Native American settlement in New Mexico, he determined that the trees were felled 800 years before Columbus. By 1936, the Colorado Plateau chronology extended back 2,000 years.

The technique pioneered by Douglass, known as crossdating, has been applied by archaeologists in many areas. In Europe, dendrochronologists working with oak trees have developed a tree-ring time line extending over the past 10,000 years. By analyzing the rings of long-dead oaks, scientists are gathering information on how the climate has changed over the past 10,000 years.

The growth rings of trees may be one of the most familiar natural records of the changing seasons, but trees are certainly not the only organisms to preserve an orderly record of the past. In the Earth's tropical seas, coral reefs transcribe a record of the daily movement of the sun as well as the seasonal changes over the course of a year.

Coral polyps, organisms that look a bit like tiny sea anemones, filter their food from the seawater and deposit calcium carbonate, the rocky substance that makes up a coral reef. Each day, coral polyps deposit a thin layer of calcium carbonate. The thickness of these daily growth layers varies with the season, changing over the course of the year.

The records preserved in coral reefs teach a valuable lesson to anyone who would learn to read the past. To read the past, you must think on a grand scale, realizing that things you take for granted have not always been so.

Back in the 1960s, paleontologist John Wells learned to read the ancient transcript preserved in the corals. In fossil corals, he found daily and annual growth rings, like those in modern corals. When he counted those daily rings, the corals revealed that the number of days in a year has not always been 365. Four hundred million years ago, corals recorded 400 days in a year. According to Wells and the fossil corals, days were shorter back then.

Physicists say you can blame that change on the moon and the tides. The tides rise and fall because the gravitational pull of the moon raises a bulge of water. The Earth rotates, pulling this bulge of water out from under the moon. But the

Creating a Time Line with Tree Rings (below)

The illustration below shows three core samples, cylinders of wood extracted from three different trees: a living tree, a standing dead tree, and a log post in an ancient building. The samples are arranged so that the most recently formed rings are on the right.

Comparing the rings, you can see that some of the rings in the living tree form a pattern matching that of some rings in the dead standing tree. These matching patterns indicate that the two trees were alive at the same time, experiencing similar growing conditions and laying down similar patterns of rings. You can also see that some of the older rings in the standing dead tree match later rings in the log post.

Each ring of a tree represents a year of growth. Starting with a living tree and working backward, you can assign a year to each ring. In this way, a dendrochronologist, a scientist who studies tree rings, can build a time line that extends back beyond the life span of any single tree. This technique is known as crossdating.

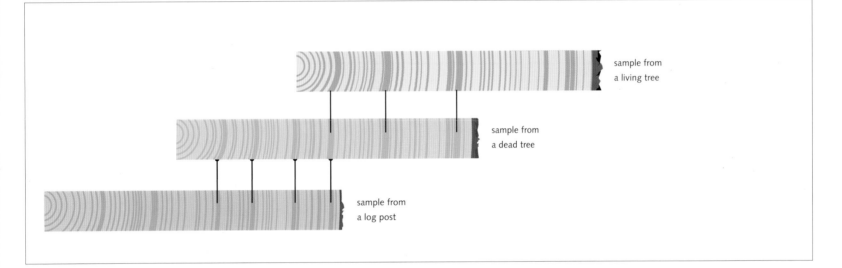

sample from a living tree

sample from a dead tree

sample from a log post

Rings in a Douglas Fir Post, Broken Flute Cave, Arizona (left)

Compare the rings of this Douglas fir, from an archaeological site in the Lukachukai Mountains of northeastern Arizona, with the rings of the tree pictured on pages 18–19. The tree on pages 18–19 shows what a dendrochronologist calls a complacent series of rings, all more or less uniform in width. This tree, on the other hand, has what a dendrochronologist calls a sensitive series, rings that vary in width with changing temperature and precipitation.

Laboratory of Tree-Ring Research, University of Arizona

California Poppies and the Flower Clock

In 1751, Swedish botanist Carl von Linne figured out how to tell time by observing the flowers in his garden. He noted what time each flower opened and closed its blossoms and divided the flowers into three groups: *Meteorici*, those that change their opening and closing times according to the weather; *Tropici*, those that change their times for opening and closing with the length of the day, and *Aequinoctales*, those that have fixed times for opening and closing.

By noticing which flowers were open, von Linne could accurately tell time, a capability that amazed watchmakers in his day. Listed among his *horologium flore*, or "flower clock," are some flowers that are still popular in modern gardens: daylilies, Icelandic poppies, and field calendula.

By carefully observing the flowers in your garden, you may be able to construct your own flower clock. You may have to watch for several days to confirm that a flower is a reliable timekeeper. Pat Murphy kept an eye on the California poppies growing near the Exploratorium and found that they close on foggy days. Rather than being *Aequinoctales*, they are *Meteorici*, changing with the weather.

Von Linne is now better known by the Latinized version of his name: Carolus Linnaeus. He went on to create what is now called the Linnean system of biological classification, in which each type of plant and animal is identified by its genus and species. His "flower clock" remains a tribute to his careful observation of the natural world.

moon continues to pull on the bulge of water, fighting the Earth's rotation. As the Earth spins, dragging the water away from the moon, the water rubs against the ocean bottom. The moon's pull on the water and the rubbing of the water against the ocean bottom slow the Earth's spin ever so gradually.

You wouldn't notice the effect of this slowing in a day, in a year, or even in a hundred years. In a hundred thousand years, the Earth slows down just enough to make the day two seconds longer. Over the course of 400 million years, the Earth's spin has slowed enough to change the day's length from twenty-two hours to its present twenty-four hours.

This tug-of-war between Earth and moon slows the Earth's spin, but it doesn't affect the time it takes the Earth to complete an orbit of the sun. The days are longer, but a year is the same length. That means the corals can record more days over the course of a year.

That's the story—according to paleontologist John Wells, according to the astronomers, and according to the record left in the corals of ancient seas. Modern astronomers have confirmed that the length of the day is increasing, measuring a change of 20-millionths of a second each year. The corals revealed it; physicists explained it; astronomers confirmed it. Deciphering the stories of nature requires people of many disciplines to share information, compare time lines, and discover the unexpected connections.

The story of a day leads to the story of a year, which leads to the story of changes taking place over millions of years. That's just the beginning.

Seasonal Changes in a Beach

Some seasonal changes are familiar; others are less so. If you are used to visiting a beach in summer, you may find the winter beach (top) shockingly small compared with the summer beach (bottom).

Every time a wave crashes against a sandy beach, it lifts some sand grains and carries them a short distance. In summer, when waves are small and lazy, they tend to carry sand grains toward the shore, forming wide beaches that slope gradually into the surf. In winter, more energetic, storm-driven waves erode the beach, pulling the sand offshore. U.S. Geological Survey

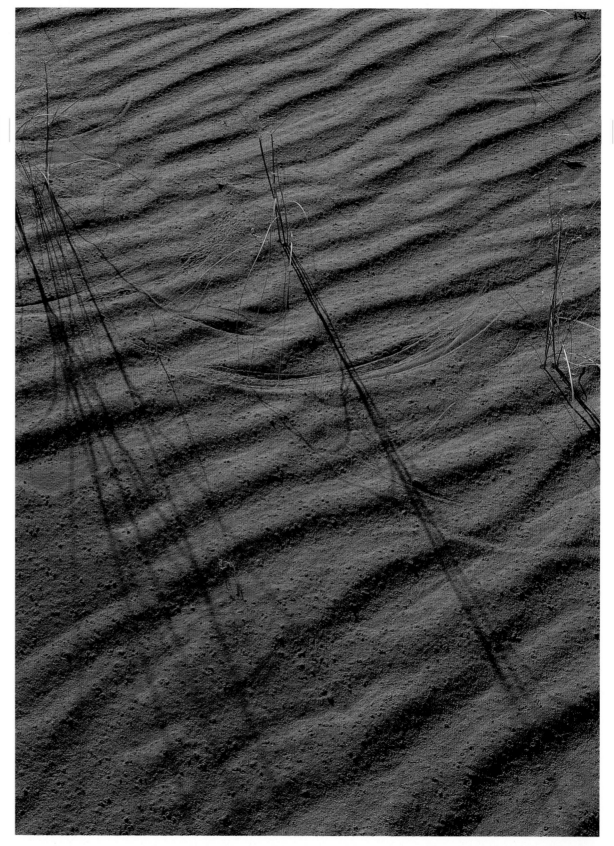

Patterns on Sand Dunes, Monument Valley Navajo Tribal Park, Arizona
Looking closely at the sand beside these blades of grass, you know that the wind has been blowing. The ripples are evidence of blowing sand— and the semicircular traces beside each plant show where moving leaves etched the sand, leaving a record of the push of a passing breeze.

28

Moving Rocks, Mojave Desert, California

On a dried lake bed in the Mojave Desert, large rocks mysteriously move from place to place, leaving tracks that provide a record of the movement. For the past fifty years, scientists have been studying the rocks and the tracks they leave behind.

Scientists don't agree on how the rocks move. Some say that the rocks freeze into ice sheets when the lake bed is flooded. The wind blows the ice sheets with their embedded rocks across the lake bed. Other scientists say that the ice sheets aren't necessary—the wind can move the rocks when the lake bed is wet and slippery. Despite years of study, no one has ever seen the rocks in motion. Experiments to determine how the rocks move are still under way.

29

THE MOON AND TIDES

"Swear not by the moon, the inconstant moon," Shakespeare's Juliet implored Romeo, fearing that his love would wax and wane like that celestial orb. But it seems unfair to call the moon "inconstant." The moon's changes are predictable, recurring with faithful regularity.

Our ancestors knew the moon's cycle better than most of us do. With electric lights available at the flick of a switch, we pay less attention to this source of nighttime illumination. When the Exploratorium tested a moon-watching activity for children with a group of families, we discovered that the parents were equally fascinated by the changing moon, a pattern many had never noticed before.

30

Full Moon,
Glacier National Park,
Montana (opposite, top)

When the moon is bright and full and round, like this moon over Citadel Mountain, sunlight is completely illuminating the side of the moon facing the Earth. This can only happen when the sun and the moon are on opposite sides of the Earth.

The full moon always rises at sunset and always sets at sunrise. Since the sun and the moon are on opposite sides of the Earth, the Earth's rotation takes the sun out of view just as the moon comes into view.

When the sun and moon are on opposite sides of the Earth, you might think that the Earth would block the sunlight, casting a shadow on the moon. That would be the case if the plane of the Earth's orbit around the sun exactly lined up with the plane of the moon's orbit around the Earth. But the moon's orbit is tipped five degrees relative to the plane of the Earth's orbit around the sun and the Earth's shadow usually misses the moon. When the moon is exactly in the plane of the Earth's orbit, the Earth's shadow covers the moon, creating a lunar eclipse.

Crescent Moon,
Arches National Park, Utah
(opposite, bottom)

The new moon is the point in the moon's cycle when the moon looks dark to those of us on Earth. Sunlight is shining on the side of the moon that we can't see. Before and after the new moon, the moon is a thin crescent.

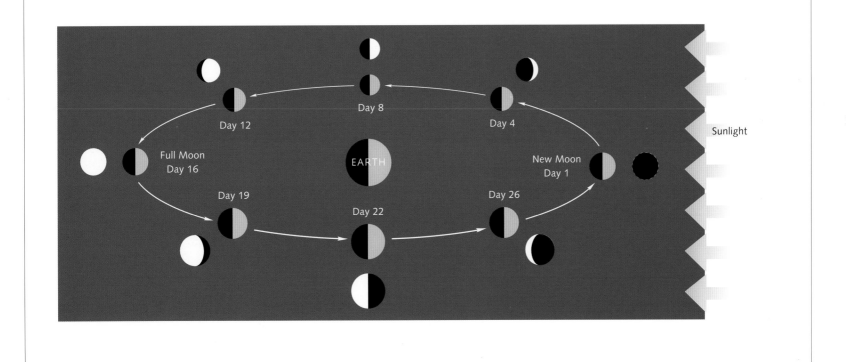

Day 8

Day 12   Day 4

Full Moon
Day 16   EARTH   New Moon
Day 1

Day 19   Day 26

Day 22

Sunlight

Phases of the Moon

The moon rises and sets with the rotation of the Earth. The moon also orbits the Earth. As the moon travels around our planet, light from the distant sun illuminates half of the moon's surface—but that illuminated surface isn't always facing the Earth. Since we see only the areas that reflect light to the Earth, the areas in darkness seem to vanish. As a result, we see a moon that seems to change shape.

When you watch the moon change shape over the course of a month, you are watching night—the lunar night—creep across the face of the moon.

Over a period of about twenty-nine days, the moon changes from new moon to full moon and back again. This diagram shows the relative positions of the Earth and moon over the course of these changes. When the side of the moon facing Earth is fully illuminated, we see a full moon.

When the side facing away from Earth is fully illuminated, the moon looks dark to us. (This diagram is not drawn to scale—to show the proper distance between the Earth and the moon, we'd need a much larger book or much smaller circles to represent the Earth and moon.)

requiring only occasional wetting by waves to survive. In a protected harbor like Santa Cruz Bay, barnacles live only as high on the rocks as the highest tide reaches. In areas with heavy surf, barnacles can live higher on the rocks, relying on splashing waves to bring them moisture and food.

Below the barnacles, the rock is hidden behind clumps of blue-black mussels and dense patches of tube worms. Mussels require more frequent wettings than barnacles, and the upper limit of the mussel bed is set by the tides, ranging up to the mean tide line, halfway between the average high and low tides.

In this photograph, a clump of common sea stars marks the lower edge of the mussel bed. Sea stars, which feed on mussels, live at lower levels than the mussels during low tide, clumping together to minimize drying. Their predation limits the invasion of mussels into the lower zones.

Toward the bottom of the photograph, the brown feather boa kelp marks the area that is largely uncovered by most low tides and covered by most high tides. Look very closely, and you'll see sea anemones wedged into crevices in the rock. Because the tide is out, they are closed and difficult to see. When the tide rises, they will open up like undersea flowers. At the very bottom of the photograph, bright green surf grass droops against the sand.

If you live by the seashore, you may have noticed a connection between the phases of the moon and the tides. The largest tidal range, the difference between the height of the low and high tides, comes at the new moon and the full moon. At these points in the moon's orbit, the sun, the Earth, and the moon are in syzygy, arranged in a line, and the sun adds its tidal pull to the tidal pull of the moon.

Tidal Zones and Rock Wall, Natural Bridges State Beach, California (above)

Various marine plants and animals make their homes on this rock wall by Santa Cruz Bay, California. This photograph was taken at low tide. The arrangement of the tide pool organisms tells an experienced observer how high the water usually rises.

Notice that the plants and animals form horizontal bands. At the top of the photograph, barnacles and a single limpet cling to bare rock. Barnacles can tolerate long exposure to air,

Bunchberry Dogwood, Acadia National Park, Maine (left)

Some seasonal changes are very familiar. Even if you don't recognize these flowers as bunchberry dogwood, you could probably guess which picture was taken in summer and which was taken in autumn. (If you need a hint, dogwood blooms in the spring and summer.)

Bunchberry dogwood spreads via subterranean stems, or rhizomes. These rhizomes develop growth rings, like the rings of a tree, allowing biologists to determine the age of these plants. Rhizomes have been found that are over fifteen feet long and thirty-six years old.

33

Autumn Forest, Baxter State Park, Maine (top right)

The colors of the trees in autumn can tell you about the weather during the preceding summer. The pigments that produce autumn color form earlier in the year, revealing themselves only when the leaves' bright green chlorophyll fades in autumn. Lighting and temperature affect the formation of these pigments, which is why you'll see the most dramatic autumn display during the cold, clear, crisp days following a warm summer.

Spring Wildflowers in Vernal Pool, Merced County, California (above)

In spring, certain areas of California grassland bloom with tiny flowers. These are vernal pools, which form where a layer of rock-hard soil, known as hardpan, prevents water from penetrating more than a few feet into the soil. In low areas, winter rains form pools. In the spring, the water evaporates and flowers bloom. These flowers are adapted to grow in flooded soils that are low in oxygen.

The best-known vernal pool species are goldfields, tiny members of the sunflower family.

The soils that produce these annual pools of gold are among the oldest in the world—up to 600,000 years old. Most soils are 20,000 years of age or younger.

Lodgepole Pines Bent by Winter Snow, Tenaya Lake, Yosemite National Park, California (above)

Notice the sharp bend in the trunk of this lodgepole pine. This kink tells you that there was a heavy snowfall when the tree was a sapling. The snow burdened the sapling almost to the breaking point, creating the bend, known as a knee, that you see here.

Lichen Showing Depth of Winter Snow, Yosemite National Park, California (right)

Even on a warm summer afternoon, the red fir forest carries a record of the winter snows. Notice the brilliant chartreuse lichen that grows on the rough bark of the trees. The lower edge of the lichen growth is about the same on every tree. That's because it marks the average height of the snowpack in winter. One of the largest fir trees in the world, the red fir thrives in very snowy locations. Red fir saplings tolerate being buried under snow until the tree is tall enough to stand above the snowpack.

Upper Yosemite Falls in Winter (left), **Spring** (middle), **and Autumn** (right)**, Yosemite National Park, California**

The waterfalls of Yosemite change with the seasons. In winter, snow dusts the cliffs near Yosemite Falls and the frozen water forms an ice cone in the pool at the base of the falls. In spring, Yosemite Falls become a torrent, swelled by the melting snow. By autumn, the falls slow to a trickle.

THE OLDEST LIVING TREES

Bristlecone pines, the oldest known living trees, grow on dry, windswept mountaintops in the western United States. Their growth rings carry a record of how the climate of this area has changed over thousands of years. Living trees provide a record that extends back almost 5,000 years. Dead trees take centuries to decay. By matching the outermost growth rings of dead pines with the inner rings of living trees, scientists have established a chronology that reaches back 9,000 years.

Core Sample of Bristlecone Pine

This piece of wood, known as a core sample, was taken by drilling a hollow shaft into the trunk of a tree to extract a cylinder of wood that shows the tree's rings. Using this method, scientists do not have to cut a tree down to examine its rings.

In this sample, you can see the rings of a bristlecone pine, dating from the late 1400s to the early 1600s. Dendrochronologists at the University of Arizona's Laboratory of Tree-Ring Research have marked this sample with tiny pinpricks. A single pinprick indicates the tenth year or decade year (1510, 1520, 1530, and so on). Two pinpricks on top of each other indicate the fiftieth year (1550, 1650, and so on); three indicate the century year (1400, 1500, and so on). Two horizontally aligned pinpricks indicate a micro-ring, a tree ring so small that it is only a few cells wide. These rings are sometimes overlooked; only crossdating will alert a dendrochronologist to look more closely for a micro-ring. Two pinpricks aligned at an angle indicate that a ring is missing from the sequence, which may happen in a year when growing conditions are particularly bad. Missing rings and other anomalies can be detected by comparing samples from several trees.

Bristlecone Pines
and Storm,
White Mountains, California

The resinous wood of the bristlecone pine does not readily decay in the high altitudes and arid conditions where the trees grow. Dead trees, like these, may remain standing for hundreds of years; fallen branches and trunks can remain intact for thousands of years.

Cross Section
of Bristlecone Pine (below)

This partial cross section was taken from the dead wood of a bristlecone pine that died thousands of years ago. The inner ring on this particular sample is from 3052 B.C. and the outermost ring is from 2642 B.C.

Where a bristlecone pine grows determines what information can be gathered from its growth rings. The growth rate of bristlecone pines is sensitive to both temperature and precipitation. Trees growing at high elevations, where lack of water does not usually limit their growth, are more sensitive to temperature variations; their rings provide a record of changing temperature. At lower elevations, lack of water limits the growth of bristlecone pines; the rings of these trees provide a record of changing precipitation.

Trunk of a Bristlecone Pine, White Mountains, California

Using the wood of bristlecone pines, scientists tested the accuracy of carbon-14 dating, a method for determining the age of organic material by analyzing the carbon it contains.

Here's how carbon-14 dating works. Nitrogen is transformed into carbon-14, or radioactive carbon, by the cosmic rays that constantly bombard the Earth's atmosphere. Plants and animals absorb radioactive carbon along with nonradioactive nitrogen as part of their metabolic processes. When a plant or animal dies, it stops absorbing carbon.

Over time, radioactive carbon decays—that is, it becomes nonradioactive nitrogen. The rate of decay is constant: half of the radioactive carbon becomes nonradioactive nitrogen every 5,730 years. William F. Libby, the physicist who developed carbon-14 dating, determined the ratio of nonradioactive and radioactive carbon in living organisms—then determined that ratio in a sample of long-dead tissue. A comparison of those ratios let him estimate the age of samples of plant or animal tissue ranging from 500 to 50,000 years old.

In the 1960s, dendrochronologists used wood of ancient bristlecone pines to test the accuracy of carbon-14 dating and discovered something that some archaeologists and physicists had suspected: the dates obtained by this method had to be adjusted. Libby's results relied on the assumption that the proportion of radioactive carbon in the Earth's atmosphere remained constant. But that isn't true. Variations in the strength of the magnetic fields of the sun and the Earth cause changes in the cosmic radiation entering the Earth's atmosphere—changes that alter the amount of radioactive carbon in the atmosphere.

Data obtained through research on bristlecone pines has been used for correcting dates obtained through carbon-14 dating. Some dates have been adjusted by over a thousand years.

38

Decay on the Forest Floor, Worcester County, Massachusetts

Notice the fungi that cover these logs, evidence that the wood has begun to decompose. This particular tree in this particular forest might have taken five to ten years to reach this stage of decomposition.

To know how long a rotten log has been lying on a particular forest floor, you need to know your forest. In tropical forests, where fungi and termites are active, a log can decompose in just five years. In the cold, wet forests of British Columbia, logs have been found fairly intact after more than 750 years.

The rate of decay also varies from species to species. A silver fir has an estimated half-life (the time it takes for half the tree to decay) of 15 years, compared with a half-life of 100 years for a western red cedar of identical size in the same environment.

In Oregon's H. J. Andrews Experimental Forest, forestry scientists have begun investigating precisely how long it takes logs to decompose, but it will be quite a while before they can report the results of their study. Back in 1985, they began an experiment designed to last 200 years.

THE PASSING CENTURIES

dies, its dead tissue will enrich the meager soil in the crack.

Eventually, as the rock weathers and crumbles, as more small plants take root, flourish, and leave their remains to fertilize the soil, a tree might manage to take root in this crack—or in the soil that plants like this one helped to create. Through the actions of lichen, mosses, grasses, wildflowers, shrubs, and trees, bare rock slowly becomes soil. It takes time—several hundred to a thousand years to make a patch of soil just a foot deep.

Pioneering plants like the stonecrop change the environment, making it more hospitable to other species. Those species, in turn, prepare the way for other plants in a process that biologists call succession, as one plant community follows and replaces another.

On the human time scale, the action of the stonecrop on the rock is imperceptible. But over the course of hundreds of years, it changes the world. In talking about events that happen over the course of hundreds of years, we are stepping just outside the time span of human experience. Few of us experienced the world a hundred years past. It's unlikely that many of us will be around to experience the world a hundred years from now. That limits our perception of the consequences of our actions.

As another example, consider the consequences of the United States' policies regarding forest management. In the 1800s, John Muir described the forests of Yosemite National Park: "The clear sunny brightness of the park is one of its most striking characteristics. Even the heaviest

Stonecrop Clinging to Cliff
(previous page)

Suppose you do something that has consequences that only become clear in a hundred years. Chances are you won't be around to observe the results of your actions—which has led Pat Murphy to joke about planting a sequoia sapling in her tiny San Francisco yard. It would be, she thinks, the world's slowest practical joke, taking hundreds of years to reveal itself as the tree reaches its full growth, dwarfing the house and filling the yard.

Joking aside, our research for this book has made us realize the power that plants have over the course of hundreds of years. In a time span just outside human experience, plants are slowly, inexorably changing the world.

We're not just talking about giant sequoias. We're talking about small, seemingly insignificant plants. Consider the cheerful sprig of stonecrop growing in a crack on page 41. Taking advantage of a bit of moisture held by some crumbled rock, this plant is making a niche for itself. Its roots are working their way deeper into the crack, releasing chemicals that form carbonic acid, which helps break down the rock. When the plant

Lichen Growth on Rock, Glacier National Park, Montana (below)

Lichen is often the first plant to grow on a bare rock surface, creeping into every tiny crack and cranny. In dry conditions, this lichen becomes desiccated and dormant. When wet by rain or dew, it absorbs water and swells up. This cycle of shrinking and swelling helps break down the rock—the lichens, which cling tightly, loosen microscopic fragments of rock as they shrink and swell. A partnership of fungi and photosynthetic algae, lichens are adapted to survive in harsh environmental conditions.

Pine Sapling Growing from Crack, Yosemite National Park, California (below)

This pine got its start when a seed fell into a crack. The seed germinated and took root in a meager soil made of rock fragments and the decaying remnants of smaller plants. Now the pine's roots are working their way into smaller cracks in the granite, pushing the cracks wider.

43

Gorge Along Avalanche Creek, Glacier National Park, Montana

In this rocky gorge, flowing water and the plants it nourishes are breaking down the bare rock. Mosses are pioneering plants; they are often among the first to grow on bare rock. Over hundreds of years, these rocks will crumble, decaying plants will add organic matter, and soil will gradually accumulate, creating an environment that will support other plants.

44

portions of the main forest belt, where the trees are tallest and stand closest, are not in the least gloomy. The sunshine falls in glory through the colossal spires and crowns. . . . The more open portions are like spacious parks, carpeted with small shrubs, or only with the fallen needles sprinkled here and there with flowers."

Less than a hundred years later, in the late 1970s, the forest no longer matched Muir's description. The vegetation had changed. In many areas, grasses and wildflowers could no longer survive on the forest floor; the canopy had grown thick, blocking the sunlight. In other areas, dense thickets of incense cedar and white fir crowded beneath the tall pines. In the groves of giant sequoias, no new saplings were growing—old trees that died left no young ones to replace them.

One simple, well-intentioned action had drastically affected the vegetation of Yosemite. Over the course of a hundred years, a very successful campaign to prevent forest fires had changed the nature of the forest it was intended to protect.

Forest fires set by lightning strikes are a natural occurrence. Before the westward migration of people of European descent, these fires burned unchecked. The growth rings of trees that survived these fires preserve a record of their frequency, as do layers of carbonized wood that sometimes remain in the soil after a fire has passed.

These natural records show that fire was a regular occurrence in Yosemite and many other forest environments. The frequency of these fires varied with the environment, but in California, an acre of ground might burn every ten to fifty years, depending on what sort of vegetation grew there.

Yosemite Forest in 1880
(top)
J. T. Boysen, courtesy of U.S. Department of the Interior, National Park Service, Yosemite National Park

Yosemite Forest in 1977
(bottom)
Compare these groves of giant sequoias. In the late 1800s, fires kept the forest floor clear of undergrowth. After a hundred years of fire suppression, the same grove is crowded with cedars and firs. By shading the forest floor, these trees prevent the growth of sequoia seedlings.
Dan Taylor, courtesy of U.S. Department of the Interior, National Park Service, Yosemite National Park

45

Under natural conditions, the plants that thrive in fire-prone areas adapt to tolerate or even require periodic burns. In Yosemite, over the century in which fires were suppressed, fire-adapted species such as ponderosa pine and black oak had begun to give way to species that could not tolerate fire, such as incense cedar and white fir. Without fires to remove this competition, the ponderosa pine and black oak were losing ground.

Fires had swept through sequoia groves every thirty to sixty years—until humans started putting these fires out. But fire—an event that humans generally regard as catastrophic—is essential to the reproduction and successful growth of these giant trees.

The bark of the mature sequoia is fire-resistant, allowing most trees to survive a fire. If the top of a sequoia is killed by fire, buds contained in the tree's roots sprout, creating a circle of shoots. Fire prepares the soil for sequoia seeds and removes shade trees like the white fir that would otherwise block the seedlings' sunlight. Only when a fire clears an area can sequoia seedlings get a start on life. If humans continue to protect sequoia groves from fire, eventually the old sequoias will die and no new seedlings will grow.

In "California's Changing Landscapes," a group of botanists and naturalists expressed the need for a 500-year management plan for the redwood forest—and in doing so, they made very clear the importance of understanding different time spans. They wrote: "Humans have the wrong life span to be natural conservationists: it is either too short or too long. If we lived only a decade, we could never develop the ability or vision to organize development projects on the scale we do now. Dam construction, large cities, highways, and complex agricultural systems would not exist because they need more time than a ten-year life span provides. If we lived 200 years we would have a long-term view of resources. But instead of ten years or 200 years, we live an average of 70: just long enough to do considerable ecological damage, and short enough to lack a conservation outlook. Of course, we can always learn to be conservationists, but that takes altruism and altered behavior that doesn't come naturally."

Developing a conservation outlook also takes an understanding of different time scales. By appreciating changes that take place over the course of hundreds of years, we can gain an understanding of the past—and a clue to where we might be going in the future.

46

Redbud and Pine, 1983 (left)

Pine, 1999 (below)

This pair of photographs shows the same two plants—a redbud tree and a pine. Since the trees are growing very close together, they are competing for limited resources—sunlight, water, nutrients in the soil. Different species of trees grow at different rates, and sixteen years later, in 1999, it is clear which tree has won the competition. The redbud has all but vanished in the shade of the pine.

FIRE AND AFTERMATH

Certain natural events—such as avalanches, forest fires, and volcanic eruptions—can wipe out all the vegetation in an area. After such a dramatic change, plant growth generally follows a natural progression that eventually restores the area to an ecological community similar to the one that was destroyed.

This natural succession is a slow process. If a wildfire burns and kills the mature trees of a forest, it may take hundreds of years for the forest to return to its prefire condition.

48

**Pine and Manzanita Growth After Fire,
Yosemite National Park,
California** (right)
In 1990, this forest burned. Nine years after the fire, the ground is covered with a thick growth of lodgepole pine saplings and manzanita shrubs.

Manzanita burns well—and recovers quickly from fire. In as little as ten to twenty days following a fire, the roots of a burned manzanita bush may send up vigorous sprouts. Fire also stimulates manzanita seeds to germinate. These seeds can survive temperatures of over 200 degrees Fahrenheit for forty minutes and still sprout.

Fire usually kills lodgepole pines, but it also prepares an ideal seedbed for new lodgepole saplings. Seeds blown in by the wind quickly take root; lodgepole pine is often one of the first trees to colonize an area following a fire.

Grasses in Burnt Lodgepole Pine Forest, Yellowstone National Park, Wyoming (left)
In the years immediately following a burn, grasses, herbs, and wildflowers sprout beneath the burned trees.

Forest Fire, Yosemite National Park, California (above)
This is a prescribed burn, a fire intentionally set by the National Park Service in Yosemite National Park. Notice that the fire is remaining on the forest floor, rather than leaping high into the treetops. Crown fires, high-intensity fires that destroy trees, are fueled in part by the buildup of dead wood and the dense growth of trees on the forest floor. Frequent burns prevent the buildup of fuel, making high-intensity fires less likely.

In the last few decades, the National Park Service and other agencies have recognized the ecological importance of fire. Today, policy stresses managing fires, rather than just putting fires out. Whenever possible, natural fires set by lightning are allowed to burn. Fires are also set so that areas can burn under controlled conditions.

Fire-Scarred Giant Sequoia, Yosemite National Park, California

The scar in the trunk of this giant sequoia documents a fire that raged through Mariposa Grove many years ago. A cross section or core sample of this tree would reveal exactly when the fire scarred the tree.

The thick bark of the mature sequoia protects these trees from all but the most intense fires—and fire is essential for the sequoia's successful reproduction. A mature sequoia can produce 300,000 to 400,000 seeds in a year, but these seeds require very specific soil conditions if they are to germinate and grow.

Fires clear away litter on the forest floor and increase the soil's ability to retain moisture, creating an ideal situation for a germinating sequoia seed. If the seedling gets enough water and enough light, it stands a chance of growing up to become one of the forest giants, growing one to two feet per year in a race to stay ahead of other trees that might shade it. A giant sequoia may grow 300 feet tall and can live to be almost 4,000 years old.

Knobcone Pine Branch, Sweetwater Point, Sierra National Forest, California

A knobcone pine reveals when the last fire passed through an area. The cones of this tree open and release seeds only when exposed to temperatures near 200 degrees Fahrenheit for about five minutes. Until that happens, the tree clings to its cones.

This knobcone pine tree sprouted after a fire consumed the previous generation of knobcone pines on this spot. The age of the tree tells you how long it has been since that fire.

Fire Scars in Cross Section of Giant Sequoia (above)

The curved lines that cut through the rings of this sequoia are fire scars. Intense forest fires killed part of the tree's cambium, the growing tissue just beneath the tree's bark. After each fire, the tree attempted to grow over the wound, a pattern of growth that created these curved rings.

The number of growth rings between fire scars shows how many years passed between intense fires. According to dendrochronologist Henri D. Grissino-Mayer, there are nineteen rings between the two fire scars in the lower left of the picture.

Notice that the tree's growth rings are wider in the years immediately following a fire. Fire burns away other trees and shrubs. Without their competition, more nutrients and water are available for the sequoia. Fire also releases nitrogen, which fertilizes the soil. Tony C. Caprio

Siesta Lake, Yosemite National Park, California (top)

Lily Pads on Siesta Lake, Yosemite National Park, California (middle)

Water Lilies and Grasses on Beaver Pond, Acadia National Park, Maine (bottom)

Lakes change slowly into meadows and meadows give way to forest in a process biologists call succession. You can see evidence of this ongoing process on the marshy shore of a California lake or a beaver pond in Maine.

Water lilies float on the lake's surface; water-loving sedges and rushes grow thick and lush in the sodden ground at the edge of the lake. When these plants die and decay, they add organic material to the lake bottom. Streams carry sediments and more organic material into the lake, contributing to the soil and helping to fill in the lake. Over time, the lake becomes shallower; the marshy land, a little drier. Eventually, grasses that require drier soil can take hold. Each successive wave of plants changes the environment subtly, making it possible for other plants to grow in this location.

You can see some of the steps in this transformation if you start at the marshy shore of a lake and move away from it. In and near the lake are plants that can live in water or tolerate marshy soil. As you move away, you find meadowland with grasses that need drier soil. Still farther from the shore, you find wildflowers and shrubs that grow at the edge of the forest. Farther still, the forest trees grow tall.

How long it takes for a lake to become a forest depends on growing conditions. At high altitudes, where conditions discourage plant growth, transformation is very slow. Under conditions more favorable to plant growth, the process progresses more rapidly. In any case, it is not a steady process. In dry years, saplings from the nearby forest may sprout in a meadow; in wet years, tree roots suffocate in the damp soil and the meadow returns.

52

Foxtail Grasses: Dating the Arrival of Non-Native Plants

When Europeans came to California, weeds such as these foxtail grasses came along with them. Botanists have figured out when certain plants were introduced by examining adobe bricks used to build California's missions. When botanists break open these mud bricks, they can find identifiable parts of plants, accidentally baked into the clay. The construction date for each mission, a matter of historical record, lets botanists determine when the plants came to California. Using this technique and others, botanists have determined that more than a thousand non-native plants have established themselves in California since 1825. In less than 200 years, the California grasslands have undergone a dramatic change.

Crashing Wave, Big Sur Coast,
California (below)

THE CHANGING COAST

The pounding surf sculpts the land. By geologists'
standards, a coast is a rapidly changing environment. You may not notice coastal changes in a
year or a decade. But over the course of hundreds
of years, the constant erosion of the crashing
waves can create arches and islands, gradually
reshaping the coastline.

54

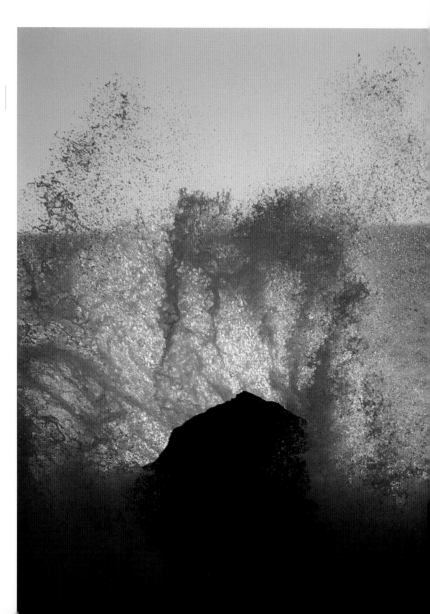

Crashing Wave, Big Sur Coast,
California (below)

Rock Archway, Big Sur Coast, California, Prior to 1980 (left)

This arch was worn in the rock by the crashing waves. A rock does not necessarily resist erosion uniformly. Parts of the rock that are softer wear away first. Fractures or cracks in the rock make some areas more susceptible to erosion, resulting in a sharp-edged rectangular opening, like this one.

An arch has a limited life span. The pounding waves that formed it will eventually destroy it.

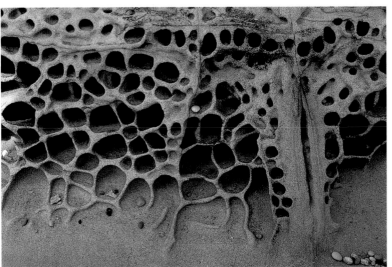 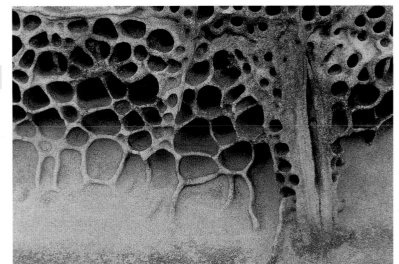

Eroded Rock, Bean Hollow State Beach, California, 1985 (above) and 1998 (above right)

Compare these photographs of the same piece of coastal sandstone, and you'll see that the patterns in the rock have changed over the years. Waves and wave-carried stones have pounded against the rock, gradually wearing it away. This erosion is most obvious in the lower left corner of the 1998 photograph, where the edges of the honeycomb cells have worn away. The changes may seem small, but small changes like these add up over the course of hundreds of years, dramatically transforming the shape of coastal rocks.

The honeycomb pattern itself is the result of erosion. The reddish lines in the rock—two vertical lines and a few horizontal lines near the top of the photograph—helped Paul Doherty figure out why the rock wore away unevenly to form honeycomb cells. The reddish lines form ridges, an indication that these lines are stronger than the surrounding rock. Paul suggests that these lines mark fractures in the original block of sandstone. Iron-rich groundwater flowed into these fractures and reacted with the rock, cementing the sand grains in the fracture together to form an area of rock that was stronger than the surrounding rock. He speculates that a similar process strengthened the rock that forms the walls of the honeycomb cells.

Sea Stack, Mendocino Coast, California (left)

This is a sea stack, an island in transition. A few hundred years ago, this rocky outcropping was part of the mainland. In another hundred years, it may be gone.

The island's shape tells you something of its origin. Notice that its rocky shores rise steeply from the water, rather than sloping gradually into the surf. This island is made of rock that has resisted the erosion of the crashing waves, remaining behind while the rock around it was worn away.

Crashing waves constantly erode the bedrock of a rocky coast, eating away at the shore. The creation of a sea stack often begins when waves wear away at fractures in the shoreline bedrock to form indentations. Parts of the rock that are softer wear away first. Over the course of years, rock that resists the pounding waves emerges as a sea stack when waves wear away the rock connecting it to the mainland.

Waves will continue to wear away this island. The life span of a sea stack depends on the rock that forms it and the energy of the waves, but it is generally measured in hundreds of years, rather than the thousands or millions of years typical of geological change.

56

Natural Bridges State Beach, Santa Cruz, California, Prior to 1980 (top right)

L. Crawford

Natural Bridges State Beach, Santa Cruz, California, 1999 (bottom right)

Over the years, erosion by the pounding surf has created a number of arches in the sandstone cliffs at Natural Bridges State Beach. Just before 1905, there were three natural bridges at this location. In 1905, one collapsed, leaving the two bridges shown in the photograph on the top right. In January 1980, after a big storm, the bridge nearest the shore collapsed. The continuing process of erosion had thinned the supporting rocks until they could no longer hold up the bridge. As the photograph on the bottom right shows, only one bridge remains. This last bridge is an island, separated from the mainland. This type of island, created by the erosion of the surrounding bedrock, is called a sea stack.

**Merced River,
Sierra National Forest,
California, 1989**
(below left)

**Merced River
After 100-Year Flood,
Sierra National Forest,
California, 1999**
(below right)

In 1997, a 100-year flood surged down the Merced River, changing the river's channel. Compare these two photographs of the river, taken ten years apart, and you'll see some changes. The undergrowth has been cleared away; boulders have been moved. The main water channel has shifted toward the camera, and a second channel has opened up on the far bank. Notice that the open region of boulders is much wider in the more recent photograph. The flood cleared the far bank of trees and carved a new river channel in the bank.

This flood was what hydrologists call a 100-year flood, but that term can be a little misleading. Since 1945, the Missouri River has had six 100-year floods, which doesn't seem to make sense.

The term 100-year flood means a flood that has a one-in-100 chance of happening in any given year. It's like having a bag of marbles in which there are 99 red marbles and one white marble. You reach into the bag and pull out a marble. You have a one-in-100 chance of pulling out the white marble each time you try—but that doesn't mean it will take you 100 tries to get the white marble. You could pull it out two times in a row, or you could try 200 times and still not get the white marble.

To predict future flooding accurately, hydrologists rely on their knowledge of past events. In 1977 one of the founding fathers of channel hydraulics, Dr. Ven T. Chow, wrote that scientists would need 100 years of records in order to predict an event that has a one-in-ten chance of happening in a given year. Currently, records of floods cover the past 100 years. To accurately predict a 100-year flood, according to Chow, hydrologists would need 1,000 years of records, which they don't have.

57

FAST ROCKS

Most rocks take tens of thousands of years to form. The rocks pictured here formed very quickly by geological standards, building up in hundreds of years, rather than thousands.

All these rocks formed from calcium carbonate, a mineral made of calcium, carbon, and oxygen. Calcium carbonate dissolves in acidic water. When water containing calcium carbonate becomes less acidic, the calcium carbonate comes out of solution, forming a solid. This chemical reaction creates rocks that tell the story of their creation—and the conditions under which they were made.

58

Tufa Formations, Mono Basin National Forest Scenic Area, California

The towering rock formations that stand in Mono Lake are the fossilized remnants of freshwater springs. These rock formations, known to geologists as tufa, always form underwater. Here, tufa towers stand well above the waterline, evidence that the lake used to be much higher than it is today.

A tufa tower grows underwater when a calcium-rich, freshwater spring seeps into briny, alkaline lake water that is rich in sodium carbonate. These dissolved chemicals interact to produce solid calcium carbonate.

Mono Lake is fed by freshwater streams, but has no outlet draining it. Water leaves the lake only by evaporation; the salts and minerals picked up by the streams remain in the lake. Mono Lake is one of the oldest continuously existing lakes in the world, estimated to be between 750,000 and 3 million years old. Over this time, Mono Lake has accumulated an estimated 280 million tons of solids in its waters. The lake's long history has resulted in a mineral content that allows rocks, like these dramatic tufa towers, to form relatively quickly.

Minerva Terrace, Mammoth Hot Springs, Yellowstone National Park, Wyoming
(top)

These pale, strangely shaped rocks reveal a history of geothermal activity at this site. The rocks are travertine, created by water rich in calcium carbonate rising from underground.

In an underground chamber, water is heated by hot magma welling up from the Earth's core. The magma expels carbon dioxide gas, which bubbles through the water, making it acidic. This hot, acidic water dissolves calcium carbonate from underground limestone.

When the hot water emerges from the earth at Mammoth Hot Springs, the carbon dioxide bubbles out and the water becomes less acidic and gradually cools off. Calcium carbonate comes out of solution and is deposited as travertine. Mats of blue-green algae aid the precipitation of the limestone. Sliced thin and viewed under a microscope, these rocks reveal the history of their creation. Their lacy layers record days when photosynthetic bacteria grew and nights when they rested.

These travertine terraces grow rapidly, building up as much as eight inches of thickness in a year. Over the course of a hundred years, these layers accumulate to create rock formations like this one.

60

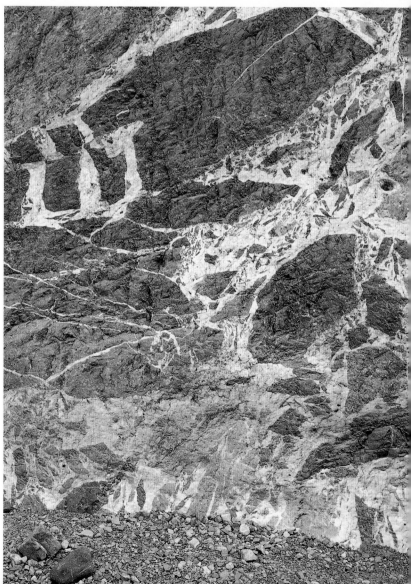

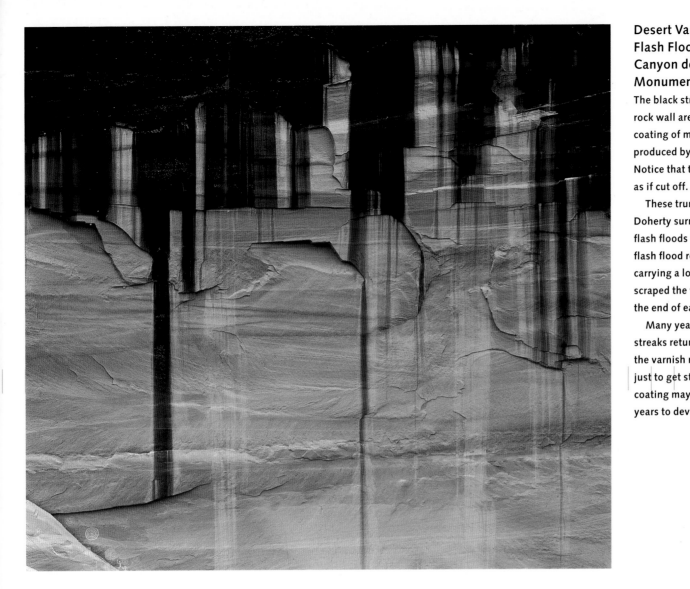

Desert Varnish Cut Off by Flash Floods, Canyon de Chelly National Monument, Arizona (left)

The black streaks that decorate this rock wall are desert varnish, a thin coating of manganese and iron oxide produced by colonies of bacteria. Notice that the streaks end abruptly, as if cut off.

These truncated streaks, Paul Doherty surmises, show a record of flash floods in this canyon. When a flash flood roared down the canyon, carrying a load of rocks and sand, it scraped the walls clean, cutting off the end of each black streak.

Many years will pass before the streaks return. According to experts, the varnish requires 80 to 100 years just to get started. A complete varnish coating may take as long as 10,000 years to develop.

61

Titus Canyon, Death Valley National Park, California

(opposite, bottom)

The walls of Titus Canyon are a mosaic pattern of dark, angular rocks, caught and held in a matrix of white stone. Unlike rounded river rocks, these dark rocks have sharp edges, indicating that they did not tumble in water long enough to become rounded. The edges of the dark rocks in the canyon wall are not resting against one another. To a geologist, this indicates that mud was caught between the rocks when they settled here, suggesting that they were carried along in a muddy slurry of water and sediment, rather than by water alone.

Geologists say that these rocks are the debris of an ancient flash flood—they broke off a larger cliff, washed down to this location, and lodged in place, all in a matter of hours. Then, over the course of a few years, water rich in calcium carbonate percolated through the sediment. Limestone deposited between the rocks cemented them in place. Over the last few hundred years, other flash floods have cut down through the deposits of this older debris flow to make the canyon with its decorated walls. The canyon is a continuing story of episodes of erosion and deposition.

TENS

OF THOUSANDS

OF YEARS ON ICE

**Glacial Polish,
Tenaya Lake,
Yosemite National Park,
California**
(previous page)

Paul Doherty likes to take people on hikes and point out interesting sights along the way. One warm summer day, he took Pat Murphy and some other Exploratorium staffers on a hike in Yosemite National Park, starting from Tenaya Lake and heading northeast.

Paul led the way over huge slabs of granite. Patches of the exposed rock shone in the light of the afternoon sun, as smooth and slick as polished marble. Scattered on the sloping granite face lay reddish boulders, also made of granite, but of a color and appearance very different from the underlying rock. "I wonder how those boulders got there," someone said.

When you are with Paul, hikes become discussions of natural history and, in certain landscapes, discussions of events in the distant past. "Those rocks?" Paul said. "They were dropped by a glacier about 12,000 years ago, give or take a few centuries. I think they came from over there." He gestured to a set of red granite peaks to the north. "I'd have to do a chemical analysis and comparison of the rock to be sure. But I know that a glacier

dropped them. I can see its footprints."

A glacier is a river of ice. The ice near the surface is brittle, like the ice that forms on lakes in the winter. But the ice deep beneath the surface, hundreds of feet down, is under pressure from the ice above. That deep ice flows, oozing like toothpaste squeezed from the tube. As the ice flows, it pulls on the brittle, overlying ice, causing it to creak and groan and snap. The flowing ice shapes the land over which it moves, leaving its tracks behind.

Throughout the rest of the hike in Yosemite, Paul pointed out the unmistakable footprints of glaciers. "Glacial polish," he said, pointing to the smooth patches of stone that glistened in the sun. Sand and sediments frozen into the moving ice had scoured the granite slopes, polishing them to a high shine. "More glacier tracks," he said, pointing to scratches and gouges in the polished granite. Rocks trapped in the ice had gouged the underlying rock. "That means the glacier was traveling along this line." He gestured along the line of the scratches.

As we walked, Paul told of other nearby locations that bore the marks of the glaciers. The mountains to the north had been sculpted by flowing ice. Nestled against the side of a mountain, a glacier can carve a steep-walled basin known as a cirque. Where two glacier-carved cirques come together, a narrow ridge called an arete forms. Where three glaciers come together, they can carve a sharp summit, known as a horn, like Matterhorn Peak. "Aretes are great fun to walk along," said Paul, who has a personal story to go with almost any geological feature. "I once

Glacial Erratics, Yosemite National Park, California (right)

On the surface of this rocky slope, you can see smooth patches of glacial polish. Glaciers buff the surfaces of underlying rock with immense pressure and a grit made of rocks that have been ground to dust. After the glacier polished this granite slab, chunks of the smooth surface eroded away, revealing rough granite that has not been subject to the glacial smoothing.

The slope is strewn with boulders of various sizes. These boulders provide more evidence that a glacier passed this way. Rocks freeze into the ice of a glacier. As the glacier slowly flows, like a gigantic river of ice, it carries the rocks along with it. When the glacier melts, it drops these boulders on the surface that it has polished smooth. Such glacier-transported rocks are called glacial erratics.

These glacial erratics are made of granite, as is the slope on which they rest. But every granite is unique in chemical composition. Analysis of these boulders would reveal that their composition doesn't match that of the slope. Glaciers can carry rocks hundreds of miles from their point of origin.

Glacial erratics come in all sizes. In England, a glacial erratic served as the site of a village—the entire town of Catworth in the county of Cambridgeshire was built atop a slab of flinty chalk left behind by a glacier.

Chart of Geologic Time (left)

Geologists divide the past into eons and eras and periods and epochs, this last being the shortest division. This chart shows the approximate starting and ending points of the eras, periods, and epochs of the last 4,600 million years. The most recent era—the Cenozoic—is divided into seven epochs, each of which ends with the syllable -cene. That syllable comes from the Greek word kainos, meaning "recent." Geologists have a funny idea of recent, by most people's standards. These seven epochs designated as "recent" span more than 65 million years.

relative duration of eras	era	period		epoch	duration in millions of years	millions of years ago
Cenozoic	Cenozoic	Quarternary		Holocene	Approx. last 10,000 years	
Mesozoic				Pleistocene	2.5	2.5
		Tertiary		Pliocene	4.5	7
Paleozoic				Miocene	19	26
				Oligocene	12	38
				Eocene	16	54
				Paleocene	11	65
	Mesozoic	Cretaceous			71	136
		Jurassic			64	190
		Triassic			35	22
	Paleozoic	Permian			55	280
		Carboniferous	Pennsylvanian		45	325
			Mississippian		20	345
Precambrian		Devonian			50	395
		Silurian			35	430
		Ordovician			70	500
		Cambrian				570
		Precambrian			4,030	

climbed up to Mathes Crest in Tuolumne. It's an arete that's a mile long with thousand-foot drops on either side. It's less than three feet wide in places. Great fun."

The very shape of Yosemite Valley is evidence that a glacier was here. A valley carved by a river usually has a V-shaped cross section. In cross section, Yosemite Valley is U-shaped, an indication that it was widened and deepened by glacial erosion. In a landscape sculpted by flowing water, tributaries usually join a river on the same level. But glacial ice carved Yosemite Valley much deeper than its tributary valleys. When the ice melted, the tributary valleys became what geologists call hanging valleys, ending high on the sides of the main valley. Streams flowing from these valleys cascade into the main valley below, forming Yosemite's famous waterfalls.

"All this leads geologists to one conclusion," Paul said. "This land was once covered with ice."

Yosemite's glaciers and the marks that they left behind, a story that Paul loves to tell, is part of a much larger story, one that encompasses tens of thousands of years and involves changes that took place all over the planet. In tens of thousands of years, as the Earth's temperature shifts and changes, the ice that fills a valley can melt, leaving a sun-warmed granite slope where chipmunks can skitter across the rocks, where we could hike comfortably in T-shirts and shorts. The story of tens of thousand of years is a tale of changing climate.

Scientists began telling this story in the early 1830s, when naturalist and geologist Louis Agassiz studied the movement of glaciers in the Swiss Alps. Agassiz saw some of the same things that

Athabasca Glacier, Jasper National Park, Alberta, Canada
Notice that these boulders don't match the rocky surface beneath them, differing in both color and texture. The boulders are glacial erratics, carried from another place by the glacier.

These boulders are beside a small stream. Though moving ice does most of the work of glacial erosion, streams on and under glaciers also contribute to the shaping of the landscape. A stream will sort sediments by size, creating sandbars, gravel bars, muddy deposits, and boulder piles. Glaciers, on the other hand, don't sort; they dump all sediments together. A glacial moraine, the accumulation of debris carried along and deposited by a glacier, contains clay, sand, and boulders all jumbled together. By looking for the sorting in a sediment pile or in a sedimentary rock, geologists can tell whether the sediments were deposited by a flowing stream or by a glacier.

Seasonal Layers in Ice, Jasper National Park, Alberta, Canada

These undulating shapes are indentations sculpted in the wall of an ice cave. The diagonal streaks are layers in the ice, marking seasonal changes in the environment.

Glacier ice is naturally layered. Each winter, fresh clean snow falls. Over time, this becomes ice, forming a white or green layer. In the summer, dirt falls on the glacier, blown by the wind or carried by avalanches, making a brown layer. These seasonal changes form yearly layers of alternating light and dark.

Geologists analyze these layers for clues about the Earth's past climate. Notice that the layers of ice in the photograph vary from year to year. A thick layer of ice indicates a year in which much snow fell and little melted.

Paul had pointed out on our hike: rock surfaces that were polished and scratched, great boulders that did not match their surroundings. Agassiz considered the rocky debris that surrounded existing glaciers and realized that these glaciers had once been much more extensive. He came up with the first version of the story that Paul told. Once upon a time, according to Agassiz's story, Europe had been engulfed in ice flowing southward from the Arctic.

Agassiz was the first to conceive of an ice age, when the planet's temperature was very different than it is today. The ice age he wrote about was a global cold period that peaked about 18,000 years ago. During that extended period of cold, glaciers covered America's Midwest and the Northeast. Rivers of ice flowed down the slopes of the Rockies and the Sierra Nevada, sculpting the land beneath them.

About 10,000 years ago, the Earth's climate warmed up and the glaciers retreated. No human observers recorded this dramatic shift in climate. A time span of 10,000 years is beyond the human experience, immeasurable on the time scales that we are accustomed to using. If you can trace your family history to the birth of your great-great-great-grandmother, you may have a record extending back some 200 years. You would have to go back five times that far to span a thousand years, fifty times that far to span 10,000. The first known written documents, the burial texts of the ancient Egyptians, date from about 4,500 B.C., only about 6,500 years ago.

The story of climatic shifts that took place tens of thousand years ago—and 10,000 years before that, and 10,000 years before that—is a detective novel, a series of deductions inferred from the evidence that those events left behind.

To piece together that story, scientists use natural environmental records. By analyzing the rings of living trees, dead standing trees, and felled trees (including ones used in human constructions), European scientists have built weather records that extend back for 10,000 years. To extend that record farther, scientists have searched for other processes that record evidence of past climate.

Appropriately enough, they found a record of past ice ages and other climatic changes preserved in ice. Areas of Antarctica and Greenland are covered with ice sheets that have been frozen for hundreds of thousands of years. Each year, snow falls onto these ice fields. The summer winds blow dirt and pollen and dust onto the ice. If snow falls in summer, as it may in the polar regions, the crystals are larger than in winter and the acid content of the snow is higher. Year after year, layers of ice build up, preserving a record of climate that extends back hundreds of thousands of years.

Like the rings of a tree, the layers of antarctic ice hold a story that scientists struggle to interpret. In *Ice Time*, a lyrical account of the new science of climate, Thomas Levenson wrote, "The ice is like a vault filled with the fragments of ancient texts, there to be reassembled by anyone with the wit to ferret them out."

In 1961, the U.S. Army contributed to the effort to read those ancient records by constructing Camp Century in Greenland, a temporary research base that housed over 200 soldiers and

Half Dome and Tenaya Canyon, Yosemite National Park, California

Half Dome, one of Yosemite's famous landmarks, is on the right side of this photograph. Tenaya Canyon runs just to the left of Half Dome. Merced River Canyon is to the right of Half Dome, just outside the frame of the photograph. Where these canyons join, they form Yosemite Valley.

The sheer cliff on Half Dome's northwestern side poses a question. It seems obvious that a piece of the dome is missing. Where did it go?

Geologists say that the missing piece is more like a quarter than a half. They say the dome was not rounded by the smoothing of a glacier, but through internal pressures in the rock. This pressure is released when sheets of rock break off the surface, creating a dome shape, in a process called exfoliation.

Geologists have found that Half Dome's cliff lines up with joints, vertical cracks that run through the bedrock deep into the Earth. This suggests that the dome was cracked along the same line as the cracks in the bedrock.

During the last ice age, a glacier quarried away the northeastern section of the dome, up to the sheer wall formed by the crack. The glacier carried those rocks away, dumping some in Yosemite Valley. Subsequent floods carried some of these rocks down to the San Joaquin Valley. There, geologists have found pieces of rock whose chemical composition matches the granite of Half Dome.

researchers. There, a team drilled down to the bedrock that lay beneath the ice, extracting cylinders of ice known as ice cores. These ice cores contained layers more than 100,000 years old.

Over the years, scientists have analyzed this ice—along with ice core samples from Antarctica and other areas of Greenland. Ice cores from some regions of Antarctica go back 400,000 years. The layers of ice have preserved any dust and debris that fell onto the ice—volcanic ash from ancient volcanic eruptions, pollen from blooming plants, salt spray from periods of very stormy weather, radioactive fallout from nuclear weapons testing.

In addition to capturing material related to specific events, the ice provides a way to estimate the Earth's temperature at different times. To estimate the temperature at the time that ice was formed, scientists analyze the oxygen atoms that make up the frozen water.

An atom of oxygen always has eight protons. But some atoms of oxygen have eight neutrons, some have nine neutrons, and some have ten neutrons. These different forms of oxygen are called isotopes. The light isotope of oxygen, with eight neutrons, and the heavier isotopes, with more than eight neutrons, are naturally present in the Earth's atmosphere, in its water, and in the ice at the poles. The ratio of the isotopes in ocean water and ice changes with shifts in the Earth's temperature. As the Earth's temperature drops, the ice that forms has more of the light isotope. When the planet is warmer, the ice has more of the heavier isotopes. Analysis of oxygen isotopes in the layers of ice has provided a record of changing climate over the past 400,000 years.

Layers of sediment that collect in lakes and oceans also preserve a record of the changing climate. In the early 1900s, Swedish geologist Gerhard De Geer began investigating the deposits that collect at the bottom of glacial lakes. These sediments, he found, vary with the seasons. In spring and summer, meltwater from the glacier carries coarse-grained sediments into the lake; during the winter, fine-grained, dark-colored sediments settle, forming a winter layer that is visually very different from the summer sediments.

De Geer called these layers varves, from the Swedish word for a periodic repetition. By counting varves, you can count back through the years. The thickness of each varve relates to the local climate—in a warm year, abundant meltwater will lay down a thick layer.

Microscopic analysis of varves offers still more information. The microorganisms contained in a given layer provide information about the climate at that time. Grains of pollen—produced by flowering plants and blown into the lake water—are well preserved in sediments. Since each species of plant produces pollen with its own distinctive shape, the pollens found in sediments reveal what plants thrived at a particular time.

Over the years, scientists have applied the principles of De Geer's work to layers of sediments in the world's oceans. Near the coast, in areas where rainy seasons alternate with dry seasons, sediments carried by rivers form seasonal varves, much like those in glacial lakes. In the deep ocean, layers of sediment are made up primarily of the skeletal remains of microscopic floating organisms called plankton. The type of plankton varies

Lembert Dome, Yosemite National Park, California
Notice that Lembert Dome has a gentle slope on one side and a steep drop-off on the other. When a glacier moves over the landscape, it smoothes the surface of rocks as it flows over them and plucks rocks from the down-flow direction, creating domes with sides that slope asymmetrically. Just one look at a landscape of these streamlined rocks screams "glacier" to a geologist.

71

with the season, and their skeletons form seasonal layers that extend back hundreds of thousands of years. By analyzing the types of plankton and the ratio of oxygen isotopes contained in their skeletons, scientists have extended the record of the Earth's changing climate back at least 700,000 years.

These natural records indicate that the last 10,000 years, the period during which human civilization has risen and flourished, has been a warm spell in the Earth's history. In the hundreds of thousands of years preceding this period, the Earth's temperature fluctuated in a series of ice ages and warm periods. Glaciers crept over the land when the temperature dropped, then retreated when the climate warmed. Jungles grew, then gave way to deserts as the climate shifted. The ocean level rose and fell as ocean water was locked into ice and as creeping glaciers pushed down on the land.

These dramatic changes left their traces on the landscape. Some of these traces—like the layers of ice and the sediments on the ocean floor—are hidden in remote locations. Others are out in the open. Any careful observer can find evidence of the last ice age in areas once covered by glaciers. Iowa farmers built fences and foundations of glacial erratics, the boulders dropped by passing glaciers. Pat Murphy remembers seeing glacial polish on rocks in the hills of New Hampshire, where she hiked as a child.

Sedimentary rocks also preserve a record of changing climate. In the American Southwest, layers of sandstone remain from when the land was covered with dunes that shifted and changed with the winds. Fossil ripples in the sandstone show which way the prevailing winds blew in that ancient time. Layers of shale made from the mud of tropical forests preserve fossils from when the desert was lush with vegetation. Limestone from deep-ocean sediments show that this land was once submerged, and oxygen isotope ratios taken from that limestone provide evidence of the temperature at that time.

Piecing together evidence from rocks, from ice, from ocean sediments, paleoclimatologists have constructed stories about what happened tens of thousands years ago—reaching back to times before the earliest humans walked the earth.

Why should we bother to consider this ancient history? In 1785, geologist James Hutton wrote "the present is the key to the past." By carefully studying natural processes in the present, he felt that scientists could understand what had happened in the past. In the case of paleoclimatology, scientists are flipping that statement around. The past, it seems, can be the key to the present. By understanding how and why the Earth's climate has shifted in the past, scientists hope to understand and predict future changes in our planet's climate—and evaluate how human activity may affect these changes.

ROCKS IN BALANCE

In the American Southwest, the bones of the Earth are laid bare. Rock formations reveal the layers of stone laid down millions of years ago— and testify to weathering of wind and water over the last 10,000 years.

Goblin Valley
State Park, Utah

A clue to the origins of these strangely shaped rock formations of Utah's Goblin Valley can be seen in the distant hills. Notice that the hills are marked by horizontal layers. These layers are different types of sedimentary rock, composed of the cemented sediments of ancient seas. The sandstone, mudstone, and siltstone of this area are rocks of different colors and, more importantly, different hardnesses.

Take a close look at the rock formations in the foreground. The caprocks, the rocks at the top of each formation, are a slightly different color than the underlying rock; the caprock and pedestal are made of different layers of stone. When a cracked layer of hard stone lies over a layer of softer stone, erosion can remove the lower layer, leaving bits of the hard layer perched on top of soft pedestals. The hard caprock protects its pedestals from erosion. Wind and rain slowly shape these rocks, sculpting them into strange configurations geologists call

hoodoos. Eventually the pedestal erodes away, dumping the caprock.

Balanced Rock, Arches National Park, Utah

Balanced Rock, a hoodoo of massive proportions, stands 128 feet tall. The huge rock at the top is estimated to weigh 3,500 tons. Gazing up at the rock, it's tempting to ask how it got up there. A more productive question might be: How did I get down here?

Study the side of Balanced Rock and the two pillars to its right, and you'll see layers of different colors. The caprock, the rock at the top of Balanced Rock, is made of slickrock sandstone, laid down in the Jurassic Period, when dinosaurs roamed the Earth. The pale tan-colored stone right below the caprock is a softer sandstone, laid down earlier. Beneath these layers is reddish Navaho sandstone, which dates to the Triassic Period.

Balanced Rock was created by the same process that created the hoodoos of Goblin Valley, operating on a grand scale. The hard caprock protects the softer sedimentary rocks of its pedestal. The pedestal erodes more slowly than the surrounding rocks, leaving the caprock perched high above the rest of the land.

Eventually the pedestal will wear through and the rock will tumble. This balanced rock used to have a companion, a smaller balanced rock that stood in the notch between it and the double neighbor rocks. That rock, which was known as Chip Off the Old Block, fell off its pedestal in 1975.

**Balancing Rocks,
Big Bend National Park,
Texas**

Eventually, the pedestal that supports
a balanced rock wears through and
the rock tumbles. This rock was once
balanced, but when the pedestal
eroded away, the rock fell onto its
neighboring rocks, where it remains
wedged.

ARCHES OF STONE

On a trip to Arches National Park, Paul Doherty visited the Fiery Furnace, a natural rock maze. The rock there has been fractured and eroded into a rectangular array of walls separated by dozens of narrow slots. "Once you are in the slots, you can't get an overview of where you are," Paul says. "All you can see is the slot you are in, and maybe one other intersecting slot. A few turns and everything begins to look alike."

The rock that makes up the maze is coral-colored sandstone, the remains of ancient desert dunes. Beneath that sandstone is a layer of salt, left by the evaporation of the sea that once covered this land. (This layer still makes local springs salty.) Because salt is less dense than sandstone, the salt rose toward the surface, lifting and cracking the overlying sandstone. Erosion widened the fractures, creating the maze of slots.

Over time, some of the weaker walls of sandstone will wear away, leaving others standing. Over the course of thousands of years, the maze might wear away to just a few sandstone walls — like the ones that became the dramatic arches of Arches National Park.

North Window,
Arches National Park,
Utah (opposite)

North Window formed in a wall of sandstone, like the ones that make up the Fiery Furnace. Look closely at the wall of rock that surrounds the window. The relatively smooth upper layer of sandstone forming the top of the window meets a layered region of rock forming the bottom. The smooth sandstone is more resistant to erosion than the layers below it. Erosion has removed the underlying layers, leaving an arch of stronger sandstone above a hole in the wall.

Look even more closely at the sandstone above the window. You can see dark vertical streaks of desert varnish, marking places where bacteria have left mineral deposits on the sandstone's surface.

Delicate Arch,
Arches National Park,
Utah

Looking at Delicate Arch, you can see evidence of what happened in the past—and what's likely to happen in the future.

Delicate Arch stands above the surrounding rock, the last bit of a sandstone wall. The arch remained when rainstorms and wind slowly eroded away the softer stone. Dark clouds behind the arch indicate that another rainstorm is on its way—and a close look at the arch suggests the consequences of continuing erosion.

Notice the smooth rock below the arch and the layered rocks that form both sides of the arch. The pillar on the left side of the arch narrows in one spot, an indication of rock more susceptible to erosion. With time, erosion will continue to wear away the sides of the arch, removing the sandstone a grain at a time. Eventually, after standing for thousands of years, the arch will fall—probably when the narrow point breaks through.

At the top of the arch, you can see more evidence of the continuing process of erosion. A line of small cavities reveals where there's a softer layer of stone, more susceptible to weathering.

Drapery Room, Mammoth Cave National Park, Kentucky

A limestone cave is a work in progress, always changing. Caves form when naturally acidic water flows through limestone rocks. When acid contacts limestone, it dissolves the rock and releases carbon dioxide. That's why a seashell bubbles if you drip lemon juice on it. A seashell is made of calcium carbonate, the same mineral that's in limestone.

Caves grow beneath the water table as groundwater moves through limestone rocks, dissolving the limestone. When the level of the water table drops, water drains out and the cave enters a new phase.

Acidic water still passes down through the cave on its way from the Earth's surface to the water table. This water dissolves limestone as it percolates through limestone rocks, becoming rich in calcium carbonate. As this calcium carbonate–rich water drips from the roof of a cave, carbon dioxide escapes, lowering the water's acidity. When the water's acidity decreases, some of the calcium carbonate in the water comes out of solution, staying on the roof of the cave in the form of calcite.

Over passing years, the calcite deposited on the ceiling by many water drops forms a stalactite. Water can also drip onto the cave's floor, forming a stalagmite. Stalactites and stalagmites can grow together to create columns. The water can also flow along walls, forming curtains and draperies that preserve ancient patterns of flowing water.

These cave formations hold a record of the past. Seasonal changes create layers in the calcite, like the rings in a tree. The precise chemical composition of the calcite is influenced by changes in climate and in the vegetation above the cave. In Iowa, Minnesota, and Wisconsin, analysis of the calcite layers in cave formations has revealed information on changes in vegetation over the last 10,000 years. Analysis of calcite layers on the walls of Devil's Hole, a cave in Nevada's Yucca Mountains, has provided information about climatic variation in the area over the last 500,000 years.

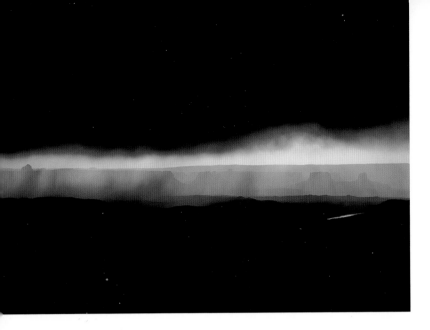

Buttes and Storm Clouds Over Green River, Canyonlands National Park, Utah (above)

The flat-topped mesas and buttes in the distance tell you about the forces that shaped this landscape. Rivers, such as the Green River (visible in the photograph as a bright streak of light), have eroded the flat sedimentary rocks of the American Southwest. These rivers rise in the snowy mountains of the Rockies and flow through the sedimentary rocks, eroding them into deep canyons. At the later stages of erosion, all that is left of a layer is the flat-topped mesas and buttes. Everything else is worn away. Desert rainstorms like this one cause flash floods and dramatic localized erosion.

Erosion Near Ubehebe Crater, Death Valley National Park, California (below)

These eroded gullies present two views of time. Look at the walls of the gullies and you can see layers of different colored sediment, deposited over hundreds of thousands of years. Consider the gullies themselves and you are looking at the action of erosion taking place over thousands of years.

On the average, Death Valley gets 2.28 inches of rain each year. Paul Doherty recalls a visit to Death Valley in which almost the full year's allotment fell in a single day—and he observed firsthand the forces that created these gullies. The water flowed across the desert landscape, gathering sediments, roaring through ditches and canyons, and enlarging the channels as he watched.

Few plants grow in the desert, so there are no plant leaves to break the falling raindrops into mist and no roots to anchor the soil. The soil is rich in clay, and the water does not soak in quickly. The drops flow downhill, joining to form a sheet of water that washes away soil to make gullies.

Geologists estimate that Ubehebe Crater formed about 6,000 years ago when steam, from water heated by hot magma, exploded through the desert floor. The explosion left a steep-sided crater. Since then, each year's rains have etched gullies in the crater's sides.

Look closely at the lines along which these granite boulders meet. Some of the boulders are separated by a gap; others by no more than a crack. Where the rocks are separated by a gap, the edge of each rock is rounded.

The slab of granite from which these boulders came formed underground. Movements of the Earth folded and stretched the granite, creating linear fractures or cracks. While the boulders were still underground, water seeped into the cracks and reacted chemically with the stone. Granite is an assemblage of minerals, which includes quartz and feldspar. Water reacts chemically with feldspar, changing it into clay and weakening the rock.

When this cracked and weakened slab of granite was pushed up to the Earth's surface, rain and wind washed away the weaker clay and chipped grains of sand from the rocks, rounding the edges. In cold weather, water froze in the cracks, expanding and forcing the cracks to widen. Over time, repeated freezing and thawing broke the chunk of cracked granite into blocks. Through these natural processes, collectively known as weathering, a thin crack gradually becomes a gap between two rounded rocks.

A CRACK IN A ROCK

Most people think of rocks as permanent, as solid as the Rock of Gibraltar. But rocks are always changing in a process that geologists call the rock cycle. Big rocks exposed to wind and weather break down into little rocks, and little rocks break down into grains of sand and the tiny particles of minerals that make up soil. Those tiny particles of sand and silt and soil are eventually washed into a lake or a sea, where they are cemented back together to make sedimentary rocks.

Next time you see a crack in a rock, take a close look. You're seeing the rock cycle in action. Stick around for a few thousand years, and you'll see that crack go through some changes.

Jumbo Rocks, Joshua Tree National Park, California
(top left)

The edges of these chunks of granite are softly rounded, worn smooth. Study the cracks in the bedrock beneath the boulders. Cracks like these were the first step in the creation of the boulders that sit atop the rock layer.

Over time, the boulders on top of the bedrock will weather away. Wind and weather will also work on cracks in the bedrock, wearing away the rock and creating separate boulders from the solid layer of rock.

The time it takes boulders like these to become sand and soil depends on the climate and the hardness of the rock. You can get a feel for rates of weathering in your area by visiting an old cemetery. Most gravestones are marked with the date that they were put in place. By examining how much gravestones have weathered since their placement, scientists have estimated weathering rates for specific rocks in specific climates.

Slot Canyon, Arizona (bottom left)

Flash floods erode the desert landscape. A crack in a rocky surface can become a slot canyon like this one, carved through rock of uniform hardness by flowing water that has nowhere else to go. Sedimentary layers in the rock create textures in the canyon walls.

Cracked Boulder, Yosemite National Park, California (below)

Look closely at the pebbles that have fallen into the crack in this boulder. In winter, water gets into cracks like this one and freezes. As water freezes, it increases in volume, pushing the crack wider. The pebbles slide deeper into the crack as it widens. When the weather warms and the ice melts, the pebbles hold the crack open, exposing the surfaces on either side of the crack to weathering.

ROUNDED ROCKS

Compare the sandstone boulders at California's Bowling Ball Beach with the rounded granite boulders on the beach at Maine's Acadia National Park. At first glance, they appear similar, but they were formed by very different processes operating on a similar time scale, taking shape over the course of tens of thousands of years.

Sandstone Concretions, Bowling Ball Beach, Mendocino State Park, California

As the tide goes out at Bowling Ball Beach, the receding surf reveals spherical rocks arranged in orderly lines, the "bowling balls" of Bowling Ball Beach.

Look carefully at the smooth rock wall behind the bowling balls, and you can see a dark circular spot. This is a bowling ball rock emerging from the wall. The spherical bowling ball is harder than the surrounding rock.

When the softer, surrounding rock gradually weathers away, the bowling ball drops to join its fellows.

The bowling balls rest on a wave-cut platform, a flat shelf eroded from the cliffs by the waves. The platform and the rocky wall behind it are made of layered sedimentary rock. Some layers are soft and easily eroded, and some are hard and resist erosion. When they formed, these layers were horizontal, but over time they have been tipped upward at a steep angle. The waves have eroded the softer layers, creating a series of grooves that parallel the coast.

When a bowling ball drops from the cliff, the waves roll it around, adding a final touch of rounding. Then they jostle the new ball until it falls into one of the grooves, lining up neatly with its fellows.

If you were to break one of these spheres, you would see internal layers of stone. These bowling balls are concretions, rocks that grow inside of layers of sediment. Often, concretions form around a fossil—minerals from the groundwater flowing through the sediment are deposited around the

**Wave-Rounded Rocks,
Otter Cliffs,
Acadia National Park,
Maine** (right)

The granite boulders on this beach
were once part of the granite cliffs
behind them. Unlike the bowling balls
of Bowling Ball Beach, these granite
boulders were sharp-edged and angular when they broke from the cliffs.
Over thousands of years, ocean waves
have rolled and tumbled these granite
fragments, rounding them like pebbles in a stream.

   Taking this photograph at sunrise
required a long exposure, which is why
the ocean waves appear as white fog.

fossil, creating a roughly spherical
area where the rock is more firmly
cemented. When the surrounding rock
weathers away, the harder sphere is
left as another one of nature's bowling balls.

ONCE UPON A TIME,

MILLIONS OF

YEARS AGO

**Volcanic Basalt, Devil's
Postpile National Monument,
California** (previous page)

Once upon a time…. That's how many stories begin. In geologists' stories, time is measured in millions of years. And in their tales, rocks shift and change, flowing like toothpaste, squeezed from a tube. The continents drift, like toy boats on a pond. In their stories, geologists tell of how mountains rise from the depths of the sea and why islands become atolls and how rocks become sand and sand becomes rocks.

Paul Doherty likes geology and likes to tell (and hear) stories. One cold January, Paul went cross-country skiing in Oregon's Wallowa Mountains. With his friend Hal, he skied some ten miles over deep powder snow to a tent-cabin out in the wilderness. Each day, they skied up a mountain, then skied down again, this being the sort of thing that Paul and his friends like to do for fun.

While skiing near Aneroid Peak, the third highest peak in the Wallowas, Paul noticed a band of rock that he recognized. Most of us recognize old friends; geologists recognize old rocks (as well as old friends). The rock Paul saw happened to be an old friend. It was a horizontal band of rock,

near the top of the mountain, twenty feet thick and broken with regular vertical fractures. Paul recognized it immediately as volcanic basalt. A similar basalt makes up Devil's Postpile National Monument in California's Sierra Nevada, a place where Paul often hikes.

The next day, Paul and his friend climbed Aneroid Peak and crossed that windswept band of rock. Looking at it closely, Paul saw further evidence that the rock had once been lava — tiny gas bubbles in the rock.

When Paul skied down Aneroid Peak that afternoon, he knew part of the story. He knew that liquid lava had poured from the Earth, forming a horizontal layer. Over the course of hundreds of years, that lava had cooled. As it cooled, it contracted and cracked, forming the vertical columns of basalt.

That evening, Paul sat by the wood stove in the tent-cabin. One of the skiers sharing the tent-cabin happened to be a local geologist, and he told Paul the rest of the story.

"You're right," the geologist said. "It's volcanic basalt. But the most interesting part is something you can't see when you ski past." That volcanic rock, the geologist explained, had formed at the Earth's equator. He knew that because he had tested the magnetic alignment of the iron atoms contained in the rock.

Rocks that contain iron carry a record of where they formed. Like tiny compass needles, iron atoms that are free to move naturally line up with the Earth's magnetic field. When molten lava cools, the iron atoms lock into position. If the rock is moved after it cools, those atoms retain their

Earth from Space

This view of the planet Earth from the *Apollo* spacecraft provides a global perspective on the moving plates that make up our planet's crust. At the top of the photograph, you can see the Red Sea, which separates Saudi Arabia from the continent of Africa.

Geologists say that Africa and Arabia were once joined. A few million years ago, they began to drift apart and created the Red Sea, an ocean in the making. As these plates continue to drift apart, the Red Sea grows wider. NASA

87

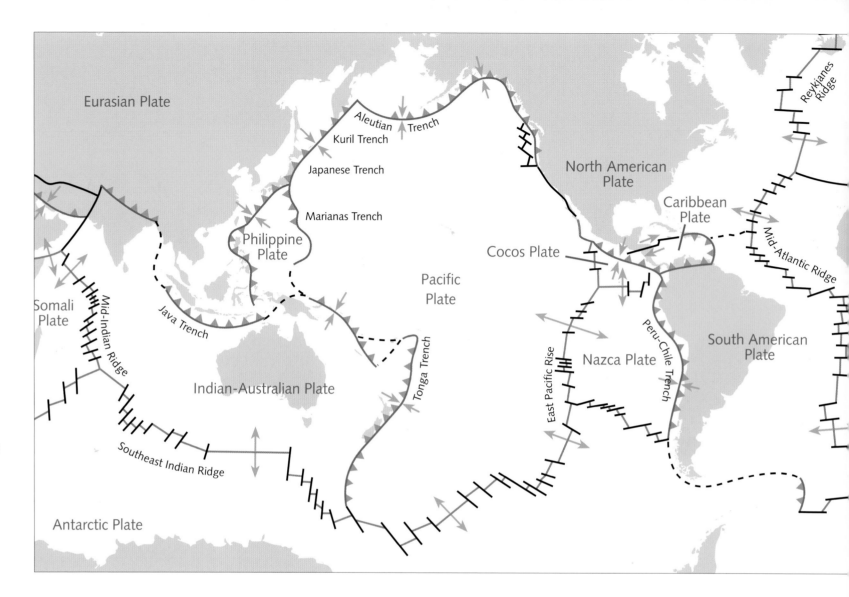

Eurasian Plate

Aleutian Trench

Kuril Trench

Japanese Trench

Marianas Trench

Philippine Plate

Java Trench

Somali Plate

Mid-Indian Ridge

Indian-Australian Plate

Southeast Indian Ridge

Antarctic Plate

Pacific Plate

Tonga Trench

East Pacific Rise

Nazca Plate

Peru-Chile Trench

North American Plate

Caribbean Plate

Cocos Plate

South American Plate

Mid-Atlantic Ridge

Reykjanes Ridge

The Earth's Tectonic Plates

The green lines on this map mark where the Earth's tectonic plates are pulling apart. The blue lines with triangles indicate where one plate is sliding beneath another. Black lines are faults, like the San Andreas fault that transects California. The Ring of Fire, an area defined by volcanoes and earthquakes, follows the edge of the Pacific Plate.

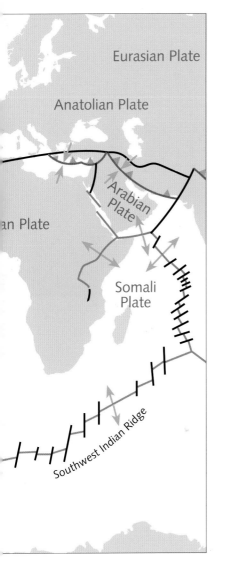

Eurasian Plate

Anatolian Plate

...an Plate

Arabian Plate

Somali Plate

Southwest Indian Ridge

original alignment, providing a record of where the rock was when it hardened.

If you were to sprinkle iron needles onto a spherical magnet (as William Gilbert, scientific experimenter and physician to Queen Elizabeth, did back in 1600), you'd find that the magnetic needles lie flat at the sphere's equator and stand on end at the poles. In between these two extreme positions, the needles assume an intermediate position.

The same is true of compass needles. At the North Pole, the north end of a compass needle dips downward, striving to stand on end. At the equator, the compass needle lies flat. The angle of a compass needle (or an iron atom) relative to the Earth's surface indicates latitude or position relative to the equator and the poles.

The iron atoms in the volcanic rocks of the Wallowa Mountains were lying flat, horizontal to the surface of the Earth, just like compass needles at the equator. To a geologist, that means that these rocks formed and cooled at the equator. That seems a little strange—how did volcanic rocks that formed at the equator end up in Oregon?

Beside the roaring fire on that cold January night, the geologist told Paul more of the story. Two hundred million years ago (give or take a few million), volcanic eruptions in the equatorial Pacific formed an arc of islands. You'd think islands would stay put—these islands formed on the equator and one would think they'd have the good sense to stay there. No such luck. Somehow, these islands migrated north and smashed into the coast of North America, forming what are now the Wallowa Mountains. The volcanic basalt that Paul

recognized as an old friend was an example of what geologists call exotic terranes, rocks that have been moved far from their point of origin.

When you are telling stories around a campfire, one story leads to another. The story of the migrating islands that became the Wallowa Mountains, with their exotic terranes, leads to a larger story that geologists like to tell, the story of plate tectonics. For the rocks of the Wallowa Mountains, that's a story that began a few hundred million years ago. For us human beings, this story began in 1620.

That was when Sir Francis Bacon noticed that the coastlines of South America and Africa fit together quite neatly, like the pieces of a puzzle. It was an intriguing observation, but one that didn't seem to lead anywhere. Back then, most people thought continents were fixed firmly in place. Rocks, after all, were solid and unchanging.

Over the passing centuries, scientists accumulated evidence that seemed to support Bacon's casual observation. Fossils revealed that a certain small reptile had thrived in the freshwater lakes of South America and Africa, an indication that those continents may have been joined. Tropical areas showed signs of ancient glaciers, which suggested that these regions were once closer to the poles. Other geological evidence suggested that the continents were once united.

In 1912, Alfred Lothar Wegener pulled together all this evidence and theorized that all the Earth's continents were once joined in a supercontinent, which he dubbed Pangea (Greek for "all earth"). Most geologists disagreed, still convinced that the continents were anchored in the underlying rock. How could they move? The whole idea was ridiculous.

When Wegener proposed his theory, geologists knew very little about the land that lay beneath the world's oceans. That changed during and after World War II, when knowledge of the oceans became essential to naval operations. During the war, many scientists aided in naval operations by gathering information about the oceans. Following the war, a serious study of the ocean floors began—and that research revealed a few surprises.

The Atlantic Ocean is bisected by a mountain range. That mountain range, known as the Mid-Atlantic Ridge, connects to another mountain range that reaches into the Indian Ocean. That mountain range, in turn, connects to another range extending into the Pacific Ocean. Altogether, a continuous chain of undersea mountain ranges, known as the oceanic ridge system, extends some 37,000 miles across the floors of the world's oceans.

Even more interesting was what geologists learned when they checked the alignment of the iron atoms in the rock on either side of these ridges. Nearest the ridge, the magnetic alignment of the rocks matched that of the Earth's magnetic field, as well they should. But in a band on either side of the ridge, the magnetic field of the rocks was the opposite of the Earth's magnetic field—all the atomic compass needles were pointing north instead of south. When geologists checked the rocks still farther from the ridge, they found bands where the magnetic alignment of the rocks once again matched the Earth's magnetic field.

How could that happen? After more investigation, geologists figured out what was going on. At the crest of each ridge, there's a crack in the Earth's crust where molten rock wells up from the Earth's hot interior. This magma cools and solidifies, adding to the Earth's crust. When the molten rock became solid, the iron atoms it contained were lined up with the Earth's magnetic field.

Geologists knew, from a variety of other evidence, that the Earth's magnetic field swapped polarity, switching north for south, at intervals ranging from a hundred thousand to a million years. Geologists had a record of when these reversals occurred. The bands of rock that had a reversed polarity must have solidified during times when the Earth's field was reversed.

The magnetic alignment of rocks on either side of the cracks revealed that the expansion of the Earth's crust had been going on for a long, long time—some 200 million years in the case of the crack that bisects the Atlantic.

All this information led geologists to revise their view of the way the world works. The continents, they decided, are not anchored in place and rocks are not really as solid as all that. The Earth is not solid all the way through, like a billiard ball. In fact, it is more like an egg, with a thin shell on the outside.

The crust of the Earth is the shell of the egg—a cracked egg. The shell is broken into a dozen large rigid plates and several smaller ones. These plates are made of rigid rock that floats on the asthenosphere, a layer of rock that's not quite solid. You can think of this rock as being sort of like Silly Putty®. Hit a blob of Silly Putty® with a hammer and it cracks; put the same blob under steady pressure, and it flows like a liquid. The rock of the asthenosphere flows very slowly, moving just an inch or so a year, but it flows.

The Earth's continents and oceans are on enormous, rigid plates that float on the asthenosphere. These plates move about, drifting away from each other and bumping into each other. The ocean ridge system marks lines where plates are moving apart, separating slowly.

In places where plates meet on or near land, their shifts and bumps create the tremors that we on the Earth's surface feel as earthquakes. In fact, geologist H. W. Menard has suggested that one way to define a tectonic plate is "a region of the crust lacking earthquakes but ringed by them." Earthquakes happen at the edges of the tectonic plates, where the Earth's crust is on the move.

On the move, yes—but that movement is too slow for us humans to perceive. We speak of things that move at "glacial speed," but glaciers are very fast indeed when compared to the Earth's crust. Each year, the Atlantic Ocean gets about two and one-quarter inches wider. Over the course of a century, the Atlantic will increase its width by about eighteen and two-thirds feet. If that sounds like a significant increase, consider the Atlantic's overall width of about 3,000 miles. This increase is about .0001, or 1/10,000, of a percent in 100 years, an easy shift to overlook.

To notice significant shifts in the Earth's crust, you'd need to watch for millions of years, a time scale so far from human perception that it's difficult to even imagine. Suppose you had a time

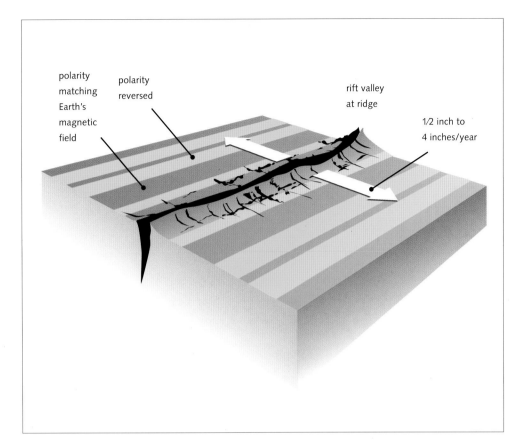

polarity matching Earth's magnetic field

polarity reversed

rift valley at ridge

1/2 inch to 4 inches/year

A Crack in the Ocean Floor

Cracks like this one top the undersea mountain ranges that mark where tectonic plates are drifting apart. Where two tectonic plates are drifting apart, a crack opens between them. This crack is filled by magma, molten rock welling up from beneath the crust. The magma cools and solidifies, then cracks again as the plates continue to move.

Iron atoms in that molten rock line up with the Earth's magnetic field, just as a compass arrow lines up with the Earth's magnetic field to point north. When the rock solidifies, the atoms lock into place, recording the magnetic field at the time the rock cooled.

Geologists know that the Earth's magnetic field swaps polarity, switching north for south, at intervals ranging from a hundred thousand to over a million years. The rocks beside these undersea cracks record these reverses in the Earth's magnetic field. Nearest the crack, where rock is currently forming, the polarity matches that of the Earth's magnetic field today. Farther from the crack, a band of rocks has the opposite polarity, dating from the last time the Earth's field re-versed, some 700,000 years ago. Still farther from the crack, another band marks the reversal before that, 1.1 million years ago.

Geologists know the approximate times of the magnetic reversals from radioactive dating. By measuring the width of these magnetic bands, they can determine how rapidly the tectonic plates are moving apart, a rate that ranges from about an inch a year at the Mid-Atlantic Ridge to just over four inches a year in the southeastern Pacific.

machine that carried you backward one year for every second that the machine was on. In one second, the duration of a human heartbeat, you would travel an entire year.

Now suppose you wanted to travel a million years back in time. For a million years to pass, at a rate of one year for each second, you'd have to turn on your time machine and travel for eleven and a half days. To travel back to see the end of the dinosaur's time on Earth, about 66 million years ago, you'd have to travel for more than two years.

To travel a billion years back in time, you'd spend a thousand times as long—11,500 days or 31½ years. At that rate, you'd spend 63 years in your time machine to reach the Precambrian Era, when the rocks of the Vishnu schist at the bottom of the Grand Canyon were forming.

The stories that geologists tell are not short stories. They are stories that take millions of years to develop and never reach an end. But the events of these stories leave traces behind—in the shape of mountains, in the layers of rock, in the sand that blows in the wind.

**Moorea,
A Young Volcanic Island**
CORBIS/Jack Fields

**Bora Bora,
An Older Volcanic Island**
(top right)
CORBIS/Yann Arthus-Bertrand

**North Male Atoll, Maldive
Islands, The Remains of a
Volcanic Island** (bottom right)
These photographs show three stages
in the development of an island.
Moorea, one of the Society Islands, is
a very young island, only about 1.5
million years old. The center of the
island is a volcanic mountain; a coral
reef has grown in the shallow coastal
waters. Bora Bora, also in the Society
Islands, is about 3 million years old,
twice the age of Moorea. The center
of the island has subsided, creating a
wider lagoon. The final stage in the
development of an island is represented by a coral atoll in the Maldive
Islands. The central mountain has
subsided beneath the ocean water,
leaving a ring of coral reefs.
CORBIS/Yann Arthus-Bertrand

Mount Everest, Tibet
(top right)

Migration of Indian Subcontinent (bottom right)

Take a close look at the peak of Mount Everest. You can see horizontal layers of rock. These layers are limestone, laid down at the bottom of the sea. In this limestone are fossils of marine organisms that thrived in the world's oceans some 50 million years ago.

Some 225 million years ago, the Indian subcontinent was a large island, separated from the rest of Asia by a vast sea. The tectonic plate that included India started drifting northward. The map on the right shows its movement. About 50 million years ago, India collided with the rest of Asia. This collision raised the seabed, creating the towering mountains of the Himalayas and the Tibetan Plateau.

Over the 30 million or so years since that initial contact, India has continued moving northward at about three and a half inches a year, shifting its position by around 1,700 miles. As this slow-motion collision continues, Mount Everest keeps on growing. Precise measurements by global positioning satellites reveal that the mountain is growing taller at a rate of about 4/100 of an inch each year.

In seventeenth-century Europe, the presence of fossil seashells on mountaintops was widely accepted as evidence of the great deluge described in the Bible. Modern geologists cite those same fossils as evidence that the plates that make up the Earth's crust drift and move over the course of millions of years.

94

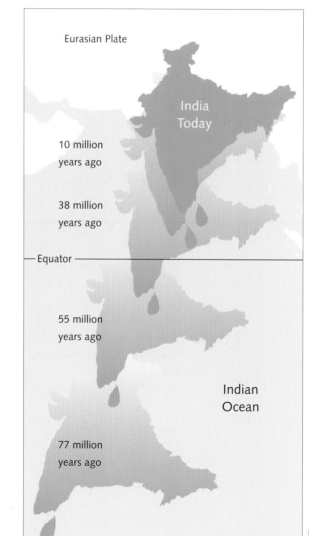

Eurasian Plate

India Today

10 million years ago

38 million years ago

Equator

55 million years ago

Indian Ocean

77 million years ago

Uplifted Strata, Stok Canyon, Himalayan Mountains, Ladakh, India (above)

In this canyon in the Himalayas, you can see layers of rock—an indication that this is sedimentary rock, formed by layers of sediment at the bottom of an ocean or lake. When sedimentary rock forms, its layers are horizontal. These originally horizontal layers are vertical, tilted on end by the forces produced by colliding continents. The alternation of soft and hard rock allows the processes of erosion to etch the landscape and reveal the layers.

Green Mountain National Forest, Vermont (top left)

Waputik Mountains, Banff National Park, Alberta, Canada (left)

The shape of a mountain reveals its age. Compare the craggy, sharp-edged slopes of the Waputik Mountains with the rounded, tree-covered slopes of the Green Mountains, part of the Appalachian range. No trees grow on the rocky slopes of the Waputiks, evidence that these mountains are still eroding. The Waputiks are young mountains, not yet worn down by the wind and weather. Old mountains, like the 400-million-year-old Appalachians, have rounded slopes blanketed with trees.

Movement of the tectonic plates can push a mountain up in 10 million years, but it takes ten times that long for that mountain's sharp edges to weather into soft folds. To understand why it takes so long to wear a mountain down, think about the last time you sat on a waterbed. When you sit

on a waterbed, your bottom sinks into the surface and the pillows at the head of the bed rise, riding higher. How high or low something floats on the surface of the bed depends on its density.

The same is true of the rocks of the Earth's crust. The Earth's crust floats on the asthenosphere, a layer of rock that's not quite solid. Mountains are made of low-density rocks; like the pillows on the waterbed, they float high on the asthenosphere, but they have roots that sink deep into the Earth. When you look at a mountain, you are seeing only about 15 percent of the whole.

As erosion wears the rock of the mountain away, the mountain floats higher on the asthenosphere, pushed up from below. When wind and weather scrape away 1,000 feet of mountain, the ground rises again, restoring 850 feet in height. To completely wear away the mountain, wind and weather must wear away the mountain and its root too.

95

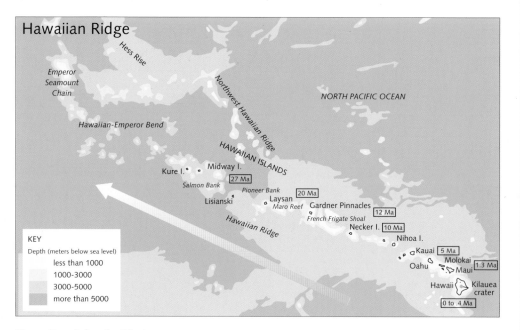

Hawaiian Ridge

Hess Rise

Emperor Seamount Chain

Northwest Hawaiian Ridge

NORTH PACIFIC OCEAN

Hawaiian-Emperor Bend

HAWAIIAN ISLANDS

Kure I.   Midway I.
27 Ma
Salmon Bank
Pioneer Bank   20 Ma
Lisianski   Laysan
Maro Reef   Gardner Pinnacles   12 Ma
Hawaiian Ridge   French Frigate Shoal
Necker I.   10 Ma
Nihoa I.
Kauai   5 Ma
Molokai
Oahu   Maui   1.3 Ma
Hawaii   Kilauea crater
0 to .4 Ma

KEY
Depth (meters below sea level)
less than 1000
1000-3000
3000-5000
more than 5000

Hawaiian Islands Chain

As the Pacific Plate moves, it leaves a trail behind—a series of islands that marks the plate's progress. This map shows the long chain of volcanic islands that includes Hawaii. The numbers in the boxes indicate the ages of the islands in millions of years (Ma). As you move from southeast to northwest along this chain, the islands increase in age. Hawaii, located at the southeastern end of the chain, is the youngest. Midway Island, halfway up the chain, is about 27 million years old. The undersea mountains at the northwest end of the chain are much older, with some estimates putting them at about 100 million years old.

The same pattern holds true in the Society Islands, the South Pacific chain that includes Tahiti. The oldest islands are at the northwest end of the chain; the youngest, at the southeast end.

In 1963, Canadian geophysicist J. Tuzo Wilson came up with a theory to explain this pattern. Beneath the Pacific Plate, he theorized, were hot spots, plumes of molten magma rising toward the Earth's surface. When a hot spot breaks through the crust, it forms a volcano. The hot spot doesn't move when the Pacific Plate shifts. As the plate moves, the hot spot burns through the crust in a new location, creating another volcano and a new island. By examining the islands and determining their ages, geologists have concluded that the Pacific Plate is moving in the direction shown by the arrow on the map.

**Lava,
Hawaii Volcanoes
National Park, Hawaii**
(above)

**Lava Flow from Pu'u O'o Vent,
Hawaii Volcanoes
National Park, Hawaii**
(left)

On Hawaii, the youngest island in the chain, active volcanoes are still adding to the island as hot lava solidifies to stone. As the Pacific Plate continues to move, these volcanoes will become dormant and new volcanic islands will form over the hot spot.

The same pattern of volcanic activity has been found on land. A hot spot beneath the North American Plate left a chain of giant volcanic craters, or calderas, extending from the Nevada-Oregon-Idaho border to the northeast up Idaho's Snake River Plain and ending at Yellowstone National Park, where the hot spot is still active. These calderas document the westward movement of the North American Plate.

CRYSTALS AND COOLING

The size of crystals within a rock can tell you something about how that rock formed.

In a crystal, atoms line up like soldiers on the march; every atom is in a precise position relative to the surrounding atoms. When lava cools slowly, the atoms have time to move around, arranging themselves into crystals. When you notice crystal grains in an igneous rock, like the salt and pepper flakes commonly found in granite, you are seeing crystals in a rock that cooled over a period of millions of years.

**Feldspar Crystals
Weathering Out of Granite,
Yosemite National Park,
California** (opposite, top)

**Glacial Erratics
with Feldspar Crystals,
Yosemite National Park,
California** (opposite, bottom)
The boulder in the foreground of the
bottom photograph on the left looks
as if it's covered with warts. In the
close-up view above it, you can see
that the warts are pale crystals of
feldspar. As the minerals surrounding
these crystals are washed away by
wind and rain, the feldspar crystals
emerge.

The feldspar crystals formed when
the granite was cooling from molten
magma. These crystals are about two
inches across. Geologists consider
crystals to be large if you can see
them with your naked eye. By those
standards, these feldspar crystals are
huge. The size of the crystals tells you
that this granite cooled slowly—it
takes long, slow cooling to produce
large crystals.

**Obsidian,
Mono Lake, California** (right)
Contrast the crystals of feldspar with
the slick surface of this obsidian.
Rather than having angular edges,
obsidian seems to flow, breaking
along curves rather than crystal faces.
It looks very much like black glass. In
fact, "volcanic glass" is another name
for obsidian. In a glass (whether it's
obsidian or the pane in your window),
the atoms are randomly arranged.
When lava cools quickly, the atoms
have no time to arrange themselves
into a crystal. If the atoms of a crystal
are soldiers on the march, the atoms
in a glass are the mob at a rock con-
cert, randomly arranged relative to
each other.

STORIES IN THE SAND

In 1785, Scottish geologist James Hutton suggested that the Earth is a sort of recycling machine. Rocks form from lava, then erode, making sand and sediment that flows to the sea. When those bits of sand or mud (along with seashells or any other sediment) are squeezed together under the right conditions, they form sedimentary rock. Sedimentary rocks are eventually pushed up from the sea to erode again.

Sand becomes sandstone and sandstone weathers to make sand. It's a never-ending cycle—from shifting sand grains to blocks of stone and back again. Each repetition of the cycle—from sand to stone and back to sand—may take 200 million years.

Cross-bedded Sandstone, Zion National Park, Utah

(opposite, top)

The patterns within a sedimentary rock—its composition and texture, the way the bits of stuff are arranged, the fossils it contains—tell geologists about how and where it formed.

Notice the pattern of fine, diagonal lines on this sandstone. These lines mark thin layers known as cross-bedding. If you were to cut through the side of a modern sand dune, you would see a similar pattern. The lines, in both sandstone and sand dune, show where the wind has sorted out layers of sand grains of different sizes. The precise pattern of the lines tells geologists about the speed and direction of the winds.

The sand grains that make up the sandstone in Zion National Park were sorted by winds that blew here hundreds of millions of years ago, when a great sand desert stretched from northern Arizona to southern Wyoming. About 65 million years ago, the Rocky Mountains rose. Rocks eroded from this mountain chain buried the sand dunes. Warmed by the Earth's interior, under pressure from the rock above, and saturated with underground water, the sand dunes became sandstone and captured the pattern of those ancient winds.

Sand Grains from Beartooth Pass, Wyoming

(opposite, bottom left)

Dave Douglass, Pasadena City College

Sand Grains from Huntington Beach, California

(opposite, bottom middle)

Dave Douglass, Pasadena City College

Sand Grains from Eureka Dunes, Death Valley National Park, California

(opposite, bottom right)

Dave Douglass, Pasadena City College

You can tell the age of a grain of sand by examining its edges. A new grain just chipped loose from a granite boulder is angular and sharp-edged, like these grains from Beartooth Pass in Wyoming. Though this sand is predominantly made of quartz, it contains feldspar, mica, and other minerals. Over time, water will react chemically with these minerals and they will wash away.

Tumbling in the wind or in moving water shapes sand grains, leaving them partially rounded, like these grains from Huntington Beach, California. These sand grains could form a layer of sediment that becomes sandstone, then weather out of the sandstone to become sand again. With each repetition of this cycle, the grains would become more rounded, until they were as rounded as the grains that make up Eureka Dunes.

Next time you're sitting on a beach or a sand dune, take a close look at the sand—but don't think of it as sand. Think of it as bits of mountains that have migrated to the sea. Or think of it as future sandstone.

Eureka Dunes, Death Valley National Park, California

Scoop up a handful of sand at Eureka Dunes, and you catch hold of the ancient past. You are holding hundreds of millions of years of history in your hand.

South Rim, Grand Canyon National Park, Arizona

When you gaze at the multicolored walls of the Grand Canyon, you are looking into the past. Here, the water of the Colorado River has laid bare the bones of the Earth, revealing 2 billion years of geological history.

A glance at the canyon walls reveals layer upon layer of rock. The most recent layers are on top; as you move downward, you move backward through time, looking at older and older layers.

Though this land is now dry, the layered stone tells of times when water laid down sediments to form sedimentary rock. Notice that some of the layers form vertical walls, while other layers have crumbled to form slopes of loose debris. Sandstone and limestone tend to break away in large pieces, forming vertical walls; the height of a wall reveals the thickness of the layer. Siltstone, mudstone, and shale crumble into smaller pieces that form gentle slopes.

The layers of rock provide an outline of grand geological events. By analyzing these layers, considering their relationships to each other, examining the rock that makes them up, and studying the fossils they contain, geologists can piece together the history of the land.

The horizontal layers of sedimentary rock date back to the Cambrian Epoch, some 500 million years ago. Below that level are rocks that have layers tilted upward. Where these rocks meet the overlying, nearly level layers, they are planed off. The contact between layerings, known as an unconformity, marks history that is missing, millions of years of the rock record removed by erosion. When you see an unconformity, you are seeing missing time.

103

Cliffs, Capitol Reef National Park, Utah (above)

The cliffs of Capitol Reef National Park have sharply defined layers that tell of events in the Triassic Period (225 to 190 million years ago). At the base of these cliffs, notice the horizontal layers, etched by the wind and weather. These layers are the sandstone, siltstone, and mudstone of the Moenkopi Formation, laid down when a shallow sea covered much of southern Utah. In this formation, geologists have found fossilized ripple marks, left by moving water millions of years ago, along with fossils of marine shellfish.

Above the red rock of the Moenkopi Formation rise the smoothly rounded rocks of the Chinle Formation. This layer is rounded, rather than fractured, because it contains bentonite clay, formed when volcanic ash decomposes. The bentonite tells geologists that volcanoes erupted and blanketed the land with ash during the Triassic.

The sheer cliffs above the layer of bentonite clay are Wingate sandstone, the remains of sand dunes that covered the area after the volcanic activity died down.

Metamorphic Rock, Merced River Canyon, California (top right)

Metamorphic Rock, Merced River Canyon, California (right)

The patterns in these rocks reveal their history. You can see layers, which indicate that the rock originally formed from sedimentary deposits. But those layers are twisted—not broken, but folded and squashed, as if the layers had become soft and malleable.

Rocks formed by the deposition of sediments can be pushed deep into the Earth's interior by the collisions of tectonic plates. Compressed by the weight of the Earth above them, warmed by the heat of the Earth's interior, layers that were once flat are compressed and twisted. These once-sedimentary rocks have been converted into metamorphic rocks by heat and pressure. The minerals in the rock have recrystallized, and the originally flat layers have been deformed.

Evidence of the Collision of Tectonic Plates, Inyo National Forest, California

In the Sierra Nevada, you can see two very different types of rock in the same location: gray granite, a rock formed when molten magma cools slowly underground, and an older, dark red metamorphic basalt. Look closely at the basalt, and you can see layers that have been warped under heat and pressure.

Some 225 million years ago, North America drifted westward, breaking away from the supercontinent of Pangea. Back then, North America was smaller—its western edge ended at Nevada. As it headed west, it bumped the Farallon Plate, a dark, dense chunk of the Earth's crust at the bottom of the Pacific Ocean. The lighter rock of North America rode up over the denser rock of the Farallon Plate, forcing the ocean plate downward into the Earth's hot interior.

Parts of the ocean plate melted in the heat, forming molten rock or magma. Lighter portions of this magma rose upward, melting their way through the crust of the Earth to form the Sierran Arc volcanoes off the coast of North America. As the North American continent continued to drift westward, it bumped into and incorporated these volcanic mountains, extending the western coast of the continent.

The Farallon Plate continued to slide beneath the continent, melting up through the crust and creating volcanic mountains like the Cascade Range. But not all the magma from the Farallon Plate reached the Earth's surface. Some pools of magma were trapped beneath the Earth where they cooled over the course of tens of millions of years. These pools formed plutons, rocks named for Pluto, the Greek god of the netherworld. Finally, just 4 million years ago, earthquakes along the Sierra-Mono Fault pushed the land to the west of the fault upward, lifting the plutons to form the granite slopes of the Sierra Nevada. The red rock that surrounds them includes volcanic and sedimentary rocks that have been changed by the heat of the cooling granite into metabasalt and metasediments.

Beach and Rock Wall, Bowling Ball Beach, Mendocino State Park, California

A glance at this rocky wall by the beach (left) reveals that it's made of layers of rock. Originally formed in horizontal layers, this sedimentary rock was tilted upward when the Pacific Plate collided with the North American Plate. Take a close look at the rock that makes up the wall (right), and you'll see patterns that swirl like ripples in water. These swirls provide a clue to the rock's formation, showing where sand and mud and clay mixed together.

Streams carry sand and mud from the land into the ocean, where they pile up on the edge of undersea canyons. Every now and then, the sediments avalanche, sending a turbulent mixture of sand and mud into the canyon. That turbulent mixing is preserved and fossilized in this sedimentary rock, known as turbidite, which always forms underwater.

Fossil Ripples Created by Algae

These ripples were made billions of years ago by the activities of cyanobacteria, also known as blue-green algae. Cyanobacteria, like green plants, are photosynthetic—that is, they can manufacture their own food using water, carbon dioxide, and sunlight. In ancient seas, the photosynthetic activity of cyanobacteria depleted the water of carbon dioxide, causing solid calcium carbonate to precipitate from the water. The calcium carbonate was trapped among the bacteria, which grew upward to form a new layer. Layer formed on layer, creating reefs like modern coral reefs. The layered fossils left by these primitive organisms are called stromatolites.

Close-up of Rock Wall, Bowling Ball Beach, Mendocino State Park, California

Fossil Worm Burrows, Bowling Ball Beach, Mendocino State Park, California

At first glance, these cracks may look like random patterns on the rock. But if you look closely, you'll notice that some of these lines wiggle through the gray rock. The sediment that fills them is a different texture and color than the rest of the rock. Those wiggling lines were once dug by burrowing worms. The empty burrows were filled in with sediment of a different color than the substrate, and the resulting burrows were fossilized and preserved in stone.

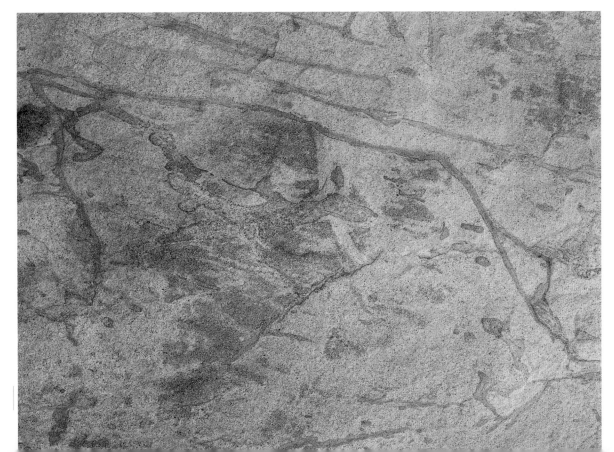

Fossil Ammonite

Fossils contained in a layer of stone can provide information on when that layer formed. The presence of a particular fossil, known as an index fossil, in two rocks found in different locations indicates that these rocks formed at about the same time, geologically speaking.

Fossil ammonites, like this one, are important index fossils. From the lower Devonian Period (almost 400 million years ago) to the upper Cretaceous (about 65 million years ago), thousands of species of ammonites evolved, thrived, and left their shells in the fossil record. At the end of the Cretaceous, ammonites became extinct. Their closest living relative is the chambered nautilus.

The shiny gold on this particular ammonite fossil comes from a naturally formed coating of iron pyrite.

Fossil Shells from Bordeaux, France, Dating from 20 Million Years Ago

On her desk, Pat Murphy has a piece of sedimentary rock in which fossil seashells are embedded. You could find better fossil shells at almost any rock shop—Pat's fossil shells are not nearly as photogenic as the ones pictured here. But Pat prefers her fossils to any that she could buy. When she was eleven years old, exploring the hills not far from the suburban development where her family lived, she found the stone containing the fossil shells in a cow pasture and marveled that those hills had once been at the bottom of the sea. It seemed then (and she says that it still seems) so unlikely that the world could change so much, that the past could be so different from the present, that traces of long ago times could linger in plain view, where anyone could just pick them up. The fossil seashells on Pat's desk have a place in her personal history, as well as in the geological history of California.

In this book, we've told you stories about how the passage of time is captured and recorded in many places—in tree rings, in ice layers, in plant communities, in stone. Now it's up to you to discover your own stories. Explore the world around you and look for places where nature has written messages, recording the past for you to read. The stories that you discover for yourself are the ones that will have the most meaning for you.

**Sunrise Reflections on Surf,
Cape Cod National Seashore,
Massachusetts**

AFTERWORD by William Neill

The Earth has many lessons to teach us. These lessons are written on the ground at your feet, on the mountain across the river from my house, in the rocks of a riverbed, in the trees on a mountainside. Once you learn to read them, you'll see them all around you.

To read this writing, one must learn the language of natural history. My academic studies of ecology, biology, and geology (in college some twenty-four years ago) gave me an overview of the many scientific disciplines that study the Earth. My knowledge of nature's language was basic then (and still is), but it was enough to give me a clear realization that the miracle of creation, this magical process of life, surrounds us every day. I was so moved by the beauty I found around me—within the ripple of ancient sediment preserved in stone or the force and energy of a fleeting thunderstorm etching the desert landscape with patterns of erosion—that I chose an expressive rather than a scientific career. I chose nature photography as my life's work in order to blend both fact and emotion on film in a personal, artistic way. The motivation for my photographic efforts was the beauty I discovered before me. The beauty that moved me most was imagery that revealed the Earth's processes.

My collection of photographs made over many years and many miles of travel betrays my fascination with the Earth. When the Exploratorium asked me to work on *Traces of Time*, we discovered that I had already photographed much of the book. I had collaborated with the Exploratorium on two other books: *By Nature's Design* and *The Color of Nature*. This time, Pat Murphy and Paul Doherty visited my studio, where we pored over images and began to build the book's visual foundation. As we looked at my images, Pat or Paul accessed the natural history described within. The stories of desert erosion, glacial epochs, and lava flows led us to more images and more stories. Not only was this a very productive process for the book — it also took me down the trail of my past explorations. As with the other book projects, I learned much more about my subjects from their stories. This added a wonderful depth to those past experiences.

After the process of editing my past work was finished, Pat and Paul selected other stories they wanted to tell and emailed their lists to me. In response, I looked further into my files or made plans to photograph the required subjects.

I also received packages in the mail from Pat and Paul. Hidden in bubble wrap were amazing treasures for me to photograph. I marveled at a fossil created by blue-green algae, life forms dating back 2.2 to 3.2 billion years. I joyfully photographed a fossil shell, entranced by its pattern and color, even though I found it odd that the sample had been painted gold. Later I learned from Paul that the shell had been fossilized by iron pyrite, better known as fool's gold. I was awed to hold in my hand a tree-ring sample from an ancient bristlecone pine, with rings dating from 3052 B.C.

Most of us know at least a few words of Earth's language. This book is designed to increase your vocabulary and heighten your awareness of the ongoing evolution of our planet. My appreciation of the Earth has been energized by working on

Traces of Time. I hope that the images and text will do the same for you, and that you will stop, now and again, to witness the miracles around us.

William Neill
Yosemite National Park, 1999
www.WilliamNeill.com

TECHNICAL NOTES

The objective of the images in *Traces of Time* is to illustrate phenomena in nature in a clear and artistic way. Many images were made before the book was conceived. These older images, dating back as far as 1982, were made with creative intentions, but not illustrative ones. Fortunately, we found many images in my archives that serve both purposes. My great challenge came in creating new images that described the subjects required by the writers and did so in an artistic way.

I have tried to apply the basic principles of good photography, which start with effective use of light and composition. I emphasize simplicity, both in my choice of equipment and in my images. I used two camera formats to photograph this book: 4x5 and 35mm. A Wista 4x5 metal field camera is my choice for most landscape photography. My lenses include 90mm, 150mm, 210mm, and 360mm focal lengths, with the 210mm and 150mm my most frequent selections. I primarily use Fujichrome Velvia Quickload film in my 4x5 camera. I enjoy the large viewing screen and slow speed that a view camera allows. I find that this

camera lends itself to a precise and contemplative approach to composing images.

For 35mm equipment, I used a Canon EOS-1N camera and an assortment of Canon lenses including the EF 20–35mm 2.8L, EF 50mm f/2.5 Macro, EF 70–200mm 2.8L, TS–E 24mm 3.5L, TS–E 90mm 2.8L, EF 28–135mm 3.5–5.6 is (Image Stabilization), and EF 300mm 4L focal lengths. A few older images were made with Nikon equipment, but the more recent images were made with Canon equipment, my 35mm system of choice.

As always on a project of such diverse subjects, some specialized equipment was required. I used my macro lens to illustrate small tree-ring and fossil samples. For these subjects, I applied a simple setup in my studio, using only natural window light and my tripod. I used Canon's Tilt/Shift lens, the 24mm and 90mm, to solve problems in depth of field. The photograph of sunflowers (page 23) is a good example. The conditions were breezy, and I was sitting on top of my van for a better angle to photograph the field of flowers. I used Canon's TS–E 90mm 2.8L with a Canon 2x Extender for the telephoto effect of heightening the density of the flowers. The tilt function allowed me to maintain great depth of field while using a fast shutter speed that would freeze the motion of the blowing flowers.

All matters of technique aside, the most important ingredient in any photograph is the photographer and his or her vision. A creative approach to photography, using a combination of practice, experimentation, immersion in the natural history of one's subject, and, most especially, a tuning in to one's own personal response to that subject, will expand any nature photographer's potential. Good luck.

113

PHOTOGRAPHY AND ILLUSTRATION CREDITS

**PHOTOGRAPHS BY
WILLIAM NEILL**

INTRODUCTION

Growth Rings of a Petrified Log—Canon EOS-1N, EF 50mm 2.5 Macro

River Rapids, Short Exposure Time—Wista 45D, Rodenstock Sironar–N 210mm f/5.6

River Rapids, Long Exposure Time—Wista 45D, Rodenstock Sironar–N 210mm f/5.6

Mount Conness and Ellery Lake, Inyo National Forest, California—Wista 45D, Schneider Apo–Symmar 150mm f/5.6

YEAR BY YEAR

Growth Rings of a Tree Cut Down in 1990—Canon EOS-1N, Canon EF 20–35mm f/2.8L

Horsetail Falls at Sunset, Yosemite National Park, California—Canon EOS-1N, Canon EF 70–200mm f/2.8L

Boulders at Sunset, 4:31 P.M.—Canon EOS-1N, Canon EF 70–200mm f/2.8L

Boulders at Sunset, 5:01 P.M.—Canon EOS-1N, Canon EF 70–200mm f/2.8L

Boulders at Sunset, 5:37 P.M.—Canon EOS-1N, Canon EF 70–200mm f/2.8L

Sunflowers and Phototropism—Canon EOS-1N, Canon TS-E 90mm f/2.8, Canon Extender EF 2.0x

California Poppies and the Flower Clock—Nikon FE, 55mm Micro-Nikkor

Patterns on Sand Dunes, Monument Valley Navajo Tribal Park, Arizona—Wista 45SP, Schneider Apo-Symmar 150mm f/5.6

Moving Rocks, Mojave Desert, California—Wista 45D, Rodenstock Sironar-N 210mm f/5.6

Full Moon, Glacier National Park, Montana—Canon EOS-1N, Canon EF 70–200mm f/2.8L, Canon Extender EF 2.0x

Crescent Moon, Arches National Park, Utah—Nikon FE2, Nikkor 80–200mm f/4.5

Tidal Zones and Rock Wall, Natural Bridges State Beach, California—Canon EOS-1N, Canon EF 70–200mm f/2.8L

Bunchberry Dogwood, Acadia National Park, Maine—Wista 45D, Schneider Apo-Symmar 150mm f/5.6

Autumn Forest, Baxter State Park, Maine—Wista 45SP, Nikkor-T*ED 360mm f/8

Spring Wildflowers in Vernal Pool, Merced County, California—Canon EOS-1N, Canon EF 20–35mm f/2.8L

Lodgepole Pines Bent by Winter Snow, Tenaya Lake, Yosemite National Park, California—Canon EOS-1N, Canon EF 70–200mm f/2.8L

Lichen Showing Depth of Winter Snow, Yosemite National Park, California—Canon EOS-1N, Canon EF 70–200mm f/2.8L

Upper Yosemite Falls in Winter, Yosemite National Park, California—Canon EOS-1N, Canon EF 70–200mm f/2.8L, Canon Extender EF 2.0x

Upper Yosemite Falls in Spring, Yosemite National Park, California—Canon EOS-1N, Canon EF 70–200mm f/2.8L

Upper Yosemite Falls in Autumn, Yosemite National Park, California—Canon EOS-1N, Canon EF 70–200mm f/2.8L

Core Sample of Bristlecone Pine—Canon EOS-1N, Canon EF 50mm Macro f/2.5, Canon Extension Tube EF25

Bristlecone Pines and Storm, White Mountains, California—Wista 45D, Rodenstock Sironar-N 210mm f/5.6

Cross Section of Bristlecone Pine—Canon EOS-1N, Canon EF 50mm Macro f/2.5, Canon Extension Tube EF25

Trunk of a Bristlecone Pine, White Mountains, California—Wista 45D, Schneider Apo-Symmar 150mm f/5.6

Decay on the Forest Floor, Worcester County, Massachusetts—Wista 45SP, Schneider Apo-Symmar 150mm f/5.6

THE PASSING CENTURIES

Stonecrop Clinging to Cliff—Wista 45SP, Rodenstock Sironar-N 210mm f/5.6

Lichen Growth on Rock, McDonald Creek, Glacier National Park, Montana—Wista 45SP, Schneider Apo-Symmar 150mm f/5.6

Pine Sapling Growing from Crack, Yosemite National Park, California—Wista 45D, Schneider Apo-Symmar 150mm f/5.6

Gorge Along Avalanche Creek, Glacier National Park, Montana—Wista 45SP, Schneider Apo-Symmar 150mm f/5.6

Redbud and Pine, 1983—Wista 45D, Rodenstock Sironar-N 210mm f/5.6

Pine, 1999—Canon EOS-1N, Canon EF 70–200mm f/2.8L

Pine and Manzanita Growth After Fire, Yosemite National Park, California—Canon EOS-1N, Canon EF 28–135mm f/3.5-5.6 IS

Grasses in Burnt Lodgepole Pine Forest, Yellowstone National Park, Wyoming—Wista 45D, Rodenstock Sironar-N 210mm f/5.6

Forest Fire, Yosemite National Park, California—Nikon FE, Tokina 28–70mm f/2.8

Fire-Scarred Giant Sequoia, Yosemite National Park, California—Wista 45SP, Schneider Apo-Symmar 150mm f/5.6

Knobcone Pine Branch, Sweetwater Point, Sierra National Forest, California—Canon EOS-1N, Canon EF 70–200mm f/2.8L, Canon Extender EF 2.0x

Siesta Lake, Yosemite National Park, California—Wista 45SP, Rodenstock Grandagon-N 90mm f/4.5

Lily Pads on Siesta Lake, Yosemite National Park, California—Wista 45D, Rodenstock Sironar-N 210mm f/5.6

Water Lilies and Grasses on Beaver Pond, Acadia National Park, Maine—Wista 45SP, Schneider Apo-Symmar 150mm f/5.6

Fox Grasses: Dating the Arrival of Non-Native Plants—Wista 45SP, Schneider Apo-Symmar 150mm f/5.6

Crashing Wave, Big Sur Coast, California—Nikon FE2, Nikkor 80–200mm f/4.5

Rock Archway, Big Sur Coast, California, Prior to 1980—Wista 45D, Rodenstock Sironar-N 210mm f/5.6

Eroded Rock, Bean Hollow State Beach, California, 1985—Wista 45D, Rodenstock Sironar-N 210mm f/5.6

Eroded Rock, Bean Hollow State Beach, California, 1998—Wista 45SP, Rodenstock Sironar-N 210mm f/5.6

Sea Stack, Mendocino Coast, California—Wista 45SP, Rodenstock Grandagon-N 90mm f/4.5

Natural Bridges State Beach, Santa Cruz, California, 1999—Canon EOS-1N, Canon EF 20–35mm f/2.8L

Merced River, Sierra National Forest, California, 1989—Wista 45D, Fujinon SW 90mm f/8

Merced River After 100-Year Flood, Sierra National Forest, California, 1999—Canon EOS-1N, Canon EF 20–35mm f/2.8L

Tufa Formations, Mono Basin National Forest Scenic Area, California—Wista 45D, Rodenstock Sironar-N 210mm f/5.6

Minerva Terrace, Mammoth Hot Springs, Yellowstone National Park, Wyoming—Wista 45D, Rodenstock Sironar-N 210mm f/5.6

Titus Canyon, Death Valley National Park, California—Wista 45SP, Schneider Apo-Symmar 150mm f/5.6

Desert Varnish Cut Off by Flash Floods, Canyon de Chelly National Monument, Arizona—Wista 45D, Rodenstock Sironar-N 210mm f/5.6

TENS OF THOUSANDS OF YEARS ON ICE

Glacial Polish, Tenaya Lake, Yosemite National Park, California—Wista 45D, Rodenstock Sironar-N 210mm f/5.6

Glacial Erratics, Yosemite National Park, California—Wista 45SP, Schneider Apo-Symmar 150mm f/5.6

Athabasca Glacier, Jasper National Park, Alberta, Canada—Canon EOS-1N, Canon EF 80–200mm f/2.8L

Seasonal Layers in Ice, Jasper National Park, Alberta, Canada—Wista 45SP, Schneider Apo-Symmar 150mm f/5.6

Half Dome and Tenaya Canyon, Yosemite National Park, California—Wista 45D, Schneider Apo-Symmar 150mm f/5.6

Lembert Dome, Yosemite National Park, California—Canon EOS-1N, Canon EF 28–135mm f/3.5–5.6 IS

Goblin Valley State Park, Utah—Wista 45D, Rodenstock Sironar-N 210mm f/5.6

Balanced Rock, Arches National Park, Utah—Wista 45SP, Schneider Apo-

Symmar 150mm f/5.6

Balancing Rocks, Big Bend National Park, Texas—Wista 45D, Schneider Apo-Symmar 150mm f/5.6

North Window, Arches National Park, Utah—Wista 45SP, Fujinon SW 90mm f/8

Delicate Arch, Arches National Park, Utah—Wista 45D, Rodenstock Sironar-N 210mm f/5.6

Drapery Room, Mammoth Cave National Park, Kentucky—Canon EOS-1N, Canon EF 20–35mm f/2.8L

Buttes and Storm Clouds Over Green River, Canyonlands National Park, Utah—Wista 45SP, Nikkor-T*ED 360mm F/8

Erosion Near Ubehebe Crater, Death Valley National Park, California—Wista 45D, Rodenstock Sironar-N 210mm f/5.6

Boulders, Alabama Hills, California—Wista 45D, Rodenstock Sironar-N 210mm f/5.6

Jumbo Rocks, Joshua Tree National Park, California—Wista 45SP, Schneider Apo-Symmar 150mm f/5.6

Slot Canyon, Arizona—Nikon FE, 24mm Nikkor

Cracked Boulder, Yosemite National Park, California—Wista 45D, Schneider Apo-Symmar 150mm f/5.6

Sandstone Concretions, Bowling Ball Beach, Mendocino State Park, California—Wista 45SP, Schneider Apo-Symmar 150mm f/5.6

Wave-Rounded Rocks, Otter Cliffs, Acadia National Park, Maine—Wista 45D, 150mm Rodenstock Sironar

ONCE UPON A TIME, MILLIONS OF YEARS AGO

Volcanic Basalt, Devil's Postpile National Monument, California—Wista 45D, Fujinon SW 90mm f/8

Mount Everest, Tibet—Nikon FE2, Nikkor 80–200mm f/4.5

Uplifted Strata, Stok Canyon, Himalayan Mountains, Ladakh, India—Wista 45D, Rodenstock Sironar-N 210mm f/5.6

Green Mountain National Forest, Vermont—Wista 45D, Fujinon SW 90mm f/8

Waputik Mountains, Banff National Park, Alberta, Canada—Wista 45SP, Schneider Apo-Symmar 150mm f/5.6

Lava, Hawaii Volcanoes National Park, Hawaii—Canon EOS-1N, Canon EF 20–35mm f/2.8L

Lava Flow from Pu'u O'o Vent, Hawaii Volcanoes National Park, Hawaii—Canon EOS-1N, Canon EF 70–200mm f/2.8L

Feldspar Crystals Weathering Out of Granite, Yosemite National Park, California—Wista 45D, Rodenstock Sironar-N 210mm f/5.6

Glacial Erratics with Feldspar Crystals, Yosemite National Park, California—Wista 45SP, Rodenstock Grandagon-N 90mm f/4.5

Obsidian, Mono Lake, California—Wista 45D, Rodenstock Sironar-N 210mm f/5.6

Cross-bedded Sandstone, Zion National Park, Utah—Wista 45D, Rodenstock Sironar-N 210mm f/5.6

Eureka Dunes, Death Valley National Park, California—Wista 45D, Rodenstock Sironar-N 210mm f/5.6

South Rim, Grand Canyon National Park, Arizona—Wista 45D, Schneider Apo-Symmar 150mm f/5.6

Cliffs, Capitol Reef National Park, Utah—Wista 45D, Rodenstock Sironar-N 210mm f/5.6

Metamorphic Rock, Merced River Canyon, California—Wista 45D, Rodenstock Sironar-N 210mm f/5.6

Metamorphic Rock, Merced River Canyon, California—Wista 45D, Rodenstock Sironar-N 210mm f/5.6

Evidence of the Collision of Tectonic Plates, Inyo National Forest, California—Canon EOS-1N, Canon EF 70–200mm f/2.8L

Beach and Rock Wall, Bowling Ball Beach, Mendocino State Park, California—Wista 45SP, Rodenstock Grandagon-N 90mm f/4.5

Fossil Ripples Created by Algae—Canon EOS-1N, Canon EF 50mm Macro f/2.5

Close-up of Rock Wall, Bowling Ball Beach, Mendocino State Park, California—Wista 45SP, Schneider Apo-Symmar 150mm f/5.6

Fossil Worm Burrows, Bowling Ball Beach, Mendocino State Park, California—Wista 45SP, Schneider Apo-Symmar 150mm f/5.6

Fossil Ammonite—Canon EOS-1N, Canon EF 50mm Macro f/2.5

Fossil Shells from Bordeaux, France, Dating from 20 Million Years Ago—Canon EOS-1N, Canon EF 50mm Macro f/2.5

ADDITIONAL PHOTOGRAPHS

Fossil Dragonfly Larva, Dating from 20 Million Years Ago—Canon EOS-1N, Canon EF 50mm Macro f/2.5

Cracked Mud on River Rocks, Green River, Utah—Wista 45D, Rodenstock Sironar-N 210mm f/5.6

Burnt Trees and Shadows on Snow, Blacktail Plateau, Yellowstone National Park, Wyoming—Wista 45D, Nikkor-T*ED 360mm f/8

Sunrise Reflections on Surf, Cape Cod National Seashore, Massachusetts—Wista 45SP, Nikkor-T*ED 360mm f/8

Ripple and Rain Drop Patterns on Sand Dunes, Monument Valley Navajo Tribal Park, Arizona—Wista 45SP, Schneider Apo-Symmar 150mm f/5.6

Close-up of River Ice—Wista 45SP, Schneider Apo-Symmar 150mm f/5.6

ADDITIONAL PHOTOGRAPHY AND ILLUSTRATION CREDITS

In addition to the photographs by William Neill, *Traces of Time* includes the work of the following photographers. All photographs remain under the copyright of the photographer except as otherwise noted.

Page 25 Laboratory of Tree-Ring Research, University of Arizona; 22, Dennis di Cicco/*Sky & Telescope: The Essential Magazine of Astronomy*; 27, U.S. Geological Survey; 45 (top), J. T. Boysen, courtesy of U.S. Department of the Interior, National Park Service, Yosemite National Park; 45 (bottom) Dan Taylor, courtesy of U.S. Department of the Interior, National Park Service, Yosemite National Park; 51 (left), Tony C. Caprio; 56 (top, right), L. Crawford (copyright © California State Parks, 1999); 87, NASA; 92, CORBIS/Jack Fields, Moorea; 93 (top and bottom), CORBIS/Yann Arthus-Bertrand; 100 (bottom left, right, and middle), Dave Douglass. The bristlecone pine core samples shown on pages 36–37 were provided by the Laboratory of Tree-Ring Research, University of Arizona.

All illustrations and maps are by Randy Comer: pages 25, 31, 65, 88, 91, 94, 96.

The map on pages 88–89 was modified from *Islands* by H. W. Menerd. Copyright 1986 by Scientific American Books, Inc. Used with permission of W. H. Freeman and Company.

Silly Putty® is a registered trademark of Binney & Smith, Inc.

Ripple and Rain Drop Patterns on Sand Dunes, Monument Valley Navajo Tribal Park, Arizona

ACKNOWLEDGMENTS

FROM PAT MURPHY AND PAUL DOHERTY

At the Exploratorium, no one works alone. This book was a collaborative effort and would not exist without the help of many members of the Exploratorium staff and friends of the Exploratorium. William Neill provided the beautiful images that inspired us. Kurt Feichtmeir kept us on track and provided much support—both moral and administrative. Rob Semper offered suggestions that helped us along the way. Megan Bury searched for images to fill in the gaps. Ruth Brown, with her keen editorial eye, helped shape the introduction. Randy Comer translated technical details into diagrams that are both visually striking and clear. Judith Dunham, as always, encouraged us when we needed it and astutely edited the material. Emanuel Gronostay arranged the workshop that helped inspire Pat Murphy to propose this project. We relied on the scientific expertise and research assistance of the following people: Charles Carlson and Norman Ten of the Exploratorium; Dr. Henri D. Grissino-Mayer of the Department of Physics, Astronomy, and Geosciences at Valdosta State University; Dr. Mark Harmon, Richardson Professor of Forest Science, Oregon State University; Dr. Rex Adams of the Laboratory of Tree-Ring Research at University of Arizona; Dr. Mark A. Schneegurt of the Department of Biological Sciences at University of Notre Dame; Dr. Dave Douglass, Department of Geology, Pasadena City College; Richard Jones of the Cambridgeshire County Council Research Group; Professor Roger Needham; and Will Crowther. And of course, none of this would have been possible without the efforts of Jay Schaefer of Chronicle Books. Thank you all for your assistance.

FROM WILLIAM NEILL

It is important for me to make note of the Exploratorium's guiding maxim, that it is "a museum of art, science and perception." The applied combination of these three words provides the forces that created this book. It is this motto that makes my work with the museum a pleasure, for the book is a collaboration that effectively heightens the reader's awareness and appreciation of this Earth. My collaborators from the Exploratorium, Pat Murphy and Paul Doherty, are gifted communicators, passionate about sharing the Earth's wonders. I thank them for the opportunity to work with them and for giving my images expanding meaning in the context of this book. Kurt Feichtmeir of the Exploratorium provided his usual steady guidance and organization through deadlines and contracts. Thanks to Jay Schaefer of Chronicle Books, and to Julia Flagg, and Scott Stowell of Open for pulling all our efforts together into a beautiful book. Diane Ackerman's wonderful foreword and Judith Dunham's editing added greatly to the project. My talented assistant, John Weller, has helped me keep abreast of my office chores while I finished this book project. I am fortunate to have his exuberance for photography and nature reinforcing my own.

Finally, I must thank my beautiful wife, Sadhna, and my daughter, Tara—Sadhna for her unfailing encouragement and support, and Tara for reminding me daily to see the world with wonder.

117

SUGGESTED READING

Close-up of River Ice

If you would like to know more about the Exploratorium, our exhibits, and our programs, check out our site on the World Wide Web. Our address is http://www.exploratorium.edu.

If you would like to know more about tree rings and dendrochronology, we recommend Dr. Henri D. Grissino-Mayer's "Ultimate Tree Ring Web Site" at http://tree.ltrr.arizona.edu/~grissino/henri.htm.

If you would like to know more about how coastlines and beaches change over time, we suggest

The Evolving Coast (Scientific American Library, 1994) by Richard A. Davis, Jr. For more on the sand that makes up those beaches, take a look at *Sand* (Scientific American Library, 1998) by Raymond Siever and at the Sand Site created and maintained by Dr. Dave Douglass at http://www.paccd.cc.ca.us/instadmn/physcidv/geol_dp/dndougla/sand/sandhp.htm.

If you are interested in how biological systems change over time, we suggest two books that describe California's ecosystems: *An Island Called California: An Ecological Introduction to Its Natural Communities* (University of California Press, 1971) by Elna Baker, and *California's Changing Landscapes: Diversity and Conservation of California Vegetation* (California Native Plant Society, 1993) by Michael Barbour, Bruce Pavlik, Frank Drysdale, and Susan Lindstrom.

To find out more about the Earth's climatic shifts and how scientists know about them, consult *Ice Time: Climate, Science, and Life on Earth* (Harper & Row, Publishers, 1989) and *Sun and Earth* (Scientific American Library, 1986) by Herbert Friedman.

If you are particularly interested in plate tectonics and the evolution of islands, consult *Islands* (Scientific American Library, 1986) by H. W. Menard. If you want to know more about the geology of specific sites in the American Southwest, we suggest the book series *Pages of Stone: Geology of Western National Parks and Monuments* (The Mountaineers, publishers) by Halka Chronic. Individual books in this series provide detailed accounts of the geologic history of the Grand Canyon and the surrounding plateaus, the Rocky

Mountains, and other areas. Another excellent series by the same author is the Roadside Geology Series, which includes *Roadside Geology of Arizona*, *Roadside Geology of Colorado*, *Roadside Geology of Utah*, and *Roadside Geology of New Mexico*. For information about the geology of Death Valley, we recommend *Geology Underfoot in Death Valley and Owens Valley* (Mountain Press Publishing Company, 1997) by Robert P. Sharp and Allen F. Glazner.

For more information on cyanobacteria, we suggest Dr. Mark A. Schneegurt's Cyanosite at http://www.cyanosite.bio.purdue.edu/index.html.

For a comprehensive but condensed view of geological time, we recommend *Geologic Time* (Prentice-Hall, Inc., 1976), a short book that covers much ground. For more general information on geology and the evolution of the landscape, we suggest *Geology Illustrated* (W. H. Freeman and Company, 1966), a wonderful book that illustrates geological concepts with drawings and photographs. For a more comprehensive approach, we recommend *The Cambridge Encyclopedia of Earth Sciences* (Crown Publishers Inc./Cambridge University Press, 1981), edited by David G. Smith.

INDEX

119